Digital and Computer Projects

Digital and Computer Projects

Robert J. Davis

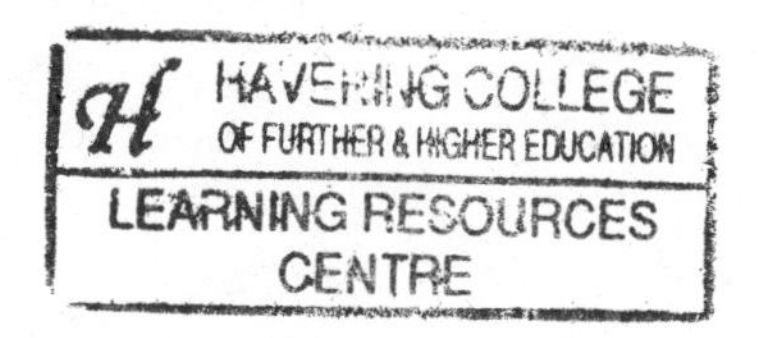

Newnes
Boston Oxford Auckland Johannesburg Melbourne New Delhi

Newnes is an imprint of Butterworth–Heinemann.

A member of the Reed Elsevier group

Recognizing the importance of preserving what has been written, Butterworth–Heinemann prints its books on acid-free paper whenever possible.

Butterworth–Heinemann supports the efforts of American Forests and the Global ReLeaf program in its campaign for the betterment of trees, forests, and our environment.

Library of Congress Cataloging-in-Publication Data

Davis, Robert J., 1956–
Digital and computer projects / Robert J. Davis.
p. cm.
Includes bibliographical references.
ISBN 0-7506-7172-6 (paperback : alk. paper)
1. Microcomputers—Amateurs' manuals. 2. Digital electronics—Amateurs' manuals. I. Title.
TK9969.D38 1999
621.39—dc21 99-12721
CIP

British Library Cataloguing-in-Publication Data

A catalogue record for this book is available from the British Library.

The publisher offers special discounts on bulk orders of this book. For information, please contact:

Manager of Special Sales
Butterworth-Heinemann
225 Wildwood Avenue
Woburn, MA 01801-2041
Tel: 781-904-2500
Fax: 781-904-2620

For information on all Butterworth–Heinemann publications available, contact our World Wide Web home page at: http://www.newnespress.com

10 9 8 7 6 5 4 3 2 1

Printed in the United States of America

Contents

Introduction

First of all, I want to give the glory to God. He is the source of all wisdom and knowledge. Second, I wish to thank my wife and family for all the countless hours I spent in tinkering and in writing this book.

This book started out as two things. First, I have always kept a book of notes on projects, etc. It contained schematics and anything I might need for frequent use. Then, I put together a series of articles on monitor conversions for *Nuts & Volts* magazine. They eventually rejected most of it.

I had people call and ask me to fax them instructions on converting monitors. Some wrote to various magazine publishers asking what to do with these surplus monitors. The responses were generally to "throw them out" and "they are not worth converting." So, I decided to put my monitor conversion plans and my project notes together into a book.

Most of the monitor conversions take 15 minutes to a couple of hours, depending on how thorough you want to be. I usually don't bother to add analog video amplifiers, because I use the converted monitors for auto telephone attendants, printer-port monitoring and control computers, and other low-key uses.

Micro Computer Journal and *Nuts & Volts* magazines have both published some articles on monitor conversion and printer port adapters from this book. I've tried to convince some of the kit makers to make some of the projects into kits. If you are interested in selling my inventions in kit form, please let me know.

This information is not guaranteed to be completely safe and accurate. The reader or builder must take all necessary safety precautions and must accept responsibility for the safety and operation of these circuits.

Who am I? I have been a hobbyist for about 30 years. I started with vacuum-tube amplifiers. I built my first home-brew computer in 1975, and I built more than 10 computers before they became commonly available. In 1980 a TRS-80 clone was almost published by *Byte* magazine. In those days I even wrote my own BIOSs. We've come a long way since then. Somehow I found time to get married, and we have three beautiful girls. I still work 12 hours a day for practically nothing. I have also developed a collection of Fox Pro and Clipper programs, most of which are available to the public domain.

Most of the projects in this book were developed for three reasons: one, just for the fun of it; two, to prove they can be easily made; and three, to save others who need them from having to "reinvent the wheel." A good rule of thumb is that if a project is no longer fun, it isn't worth it. Take a break and come back another day to finish it.

This is the fourth revision . . . the best yet, for publication by Newnes. The second revision deleted some out-of-date circuits such as the 9-IC Apple IIE MMU emulator. Apple's claim that it took more than 90 ICs was just to scare you away. The third revision added more monitors and more printer port adapters. This fourth revision adds many new inventions never before published. Among them is the 68HC11 universal or EMS (Energy Management System) controller, the 27010 (Pentium BIOS) copier, and a super-simple VGA-to-CGA converter.

To allow for growth while keeping the table of contents up-to-date, this book uses a project numbering system that makes it easier to add more projects: the chapter number, followed by a dash, and then the number of the project within that chapter. These chapters have grown substantially over the years. I keep the original in a loose-leaf notebook and add something new almost every week. If you find you enjoy making projects, you'll want to start a notebook of your own to keep all of your inventions in.

This book is written in the hope that those learning computers, electronics, and digital circuit design can have something more fun to build than a "walking ring counter." That was the most complex project required when I was in school.

My cousin John, who teaches computers and electronics, tells me that his students are looking for fun projects that combine the two areas of learning. My hope is that this book will be utilized by such schools and offer fun and excitement to the students.

Assembly Hints

Most of the designs are built using wire wrap. Wire wrap uses special IC sockets with extra-long leads to allow wires to be wrapped around it. This makes for easy assembly, modification, and disassembly. First solder in the sockets and small components. When using circuit boards without power bus runs, use 18- to 22-gauge solid copper wires to create the power and ground buses. On digital circuits add 0.01 to 0.1 μF capacitors across the buses for every one or two ICs.

Use more filter capacitors for each power source for analog ICs. Try to design the buses to come together near their point of entry onto the circuit board.

Some of the 24-pin ICs are available only in 0.3-inch-wide versions. Sockets for this size are not yet readily available. To get around that, you can use two 14-pin sockets. Remove the last two pins, then sand them off to fit the sockets end-to-end, for the 24-pin chips. Another option is to take a normal 24-pin socket and cut out the middle. This works best with sockets that only have two or three small plastic pieces to set the socket width. It is also possible to purchase strips of pins that can be arranged to make any style of socket you might need. I have seen some designs where all of the analog components are put in sockets in this manner.

Circuits shown in this book quite often have a 7805 or 7812 5- or 12-volt regulator for their power source. Use small electrolyte capacitors at the input and output of the regulator. A value of 50 to 220 μF at 16 to 25 volts works fine. When the power circuits are complete, test the power distribution with a voltmeter. When the power is safe, you can then add the ICs.

For wire-wrap wire, precut and strip several wires at 3-, 4-, and 5-inch lengths. Wrap one IC at a time until each one is completed. Then test your circuit with a logic probe or o'scope. A dual-trace 10- to 20-MHz scope is almost a necessity. A used one can be bought at ham shows for as little as $150.

Power supplies are usually just an AC adapter with 9 volts at 0.5 amps output. This is then regulated to 5 volts if needed. Some CMOS circuits don't need regulators. If a power supply is in the same box as the circuit, a metal-shielded 12.6-volt center tapped 1-amp transformer is usually used.

A regulator is needed if the schematic requires 5 volts. Sometimes I just used an AC adapter that has 5- and 12-volt regulators built in. These regulated AC adapters were formerly used in some older computer games and are most likely no longer available.

Safety Tips

When building projects, use AC adapters whenever possible. They provide isolation from the power line and hence a great deal of safety. The use of a disconnect jack is also a good idea, and the jack will reduce wear and tear on the cord.

When a power transformer must be used because of the voltages required by the project, run the 110-volt cord as directly as possible to the transformer. If it goes through a switch, use heat-shrink tubing to provide insulation and reduce the risk of shock.

Always wear shoes and keep away from wet floors. My shop is in the basement, so I keep some leftover pieces of carpet on the floor to provide extra insulation. Always be sure to check the power supply for correct operation and voltage before installing the ICs.

When working with monitors, remember there are very high voltages present. The large red wire going to the CRT (cathode ray tube) has 24,000 volts or more on the connecting clip! This is the same voltage that is used with an electric chair. Fortunately, the current is lower than that used with an electric chair. I have seen such a high voltage jump to a screwdriver blade, then over the screwdriver's handle and into my hand. (On an unsafe note: that screwdriver ended up embedded in the wall behind me.) A special high-voltage probe is required to measure such voltages.

The horizontal output transistor and deflection yoke also have about 1,000 volts on them. If safety precautions are not followed, this voltage will, at the very least, leave a nasty burn on you and destroy a voltmeter. Always turn the monitor off and unplug it before touching anything. Then, let it wait for a few minutes before touching any of the power or horizontal output devices.

If at all possible, avoid disconnecting the high-voltage lead. Sometimes in order to get to the circuit board, you have no choice but to disconnect it. Make sure the power is off and disconnected before proceeding. In order to disconnect the high-voltage lead, it must first be grounded by connecting a jumper wire, with alligator clips, to a screwdriver and to ground. Then, carefully push the screwdriver up under the high-voltage seal at the picture tube. Make sure the alligator clip is not going to slide off the screwdriver or off of the ground connection!

Quick BASIC

The source listings are in Microsoft Quick BASIC 4.5. Some of them have been modified to work with Clipper and Fox Pro. Modifying them for Visual BASIC or other versions of BASIC should not be too difficult. I prefer Quick BASIC because it makes fairly fast executables, which is necessary for them to run on old surplus computers and laptops.

Most versions of DOS come with some form of BASIC. That version is usually slower and cannot make a stand-alone (executable) program. To use DOS BASIC, exit to a DOS prompt and change to the DOS directory. It should be either Qbasic or Basica. Type "Qb" or "Basica" and press Enter to start it.

If you cannot get the source code on a disk, it can be typed in. The first two or three lines in a program that start with quotation marks can be skipped—they are comments. Also, some lines have comments in them after the code, and these can be skipped as well. They also start with quotation marks.

The software in this book is available for downloading at elim.edu/tech. Software and hardware updates will also be posted at this site.

Quick BASIC is easy to use. Just type "QB" and the name of the program to run or modify. It also has very good online help, samples, and reference information. Best of all, it works under DOS 3 and higher, so it can be used on some of the oldest computers.

I've even used Quick BASIC under Windows 95 with no problems. Multiple applications can be run on one computer using Windows. Just start the application and press Alt and Enter to make it run in a window. Then, start another application, and do the same to make it run in a window. If the computer fails to multitask, create a pif file and start the program by running the pif file instead of the exe file. Using this technique, several applications such as the printer port telephone line monitor, the printer port system clock, and others can be run on the same computer.

Making Printed Circuit Boards

Some of the projects in this book have printed circuit board designs. These were designed in Turbocad, an inexpensive CAD program for PC-compatible computers, and are shown as though you're seeing through the board from the top side. This view is for use with transfer films available from several sources. These films look like the clear plastic used with overhead projectors. One company does color its product blue.

To use a circuit board transfer film, you must first get the design copied onto the film, using a laser printer or a photocopier. Then lay the film, design side down, onto a copper clad board. Then lay a plain piece of paper over that and iron it. The correct temperature and time instructions come with the transfer film. It usually takes about 15 to 20 minutes on "cotton." The iron must keep moving, to prevent overheating or burning. The board will warp a little, making it difficult to get heat to the edges.

Then, when you've finished ironing, peel off the transfer film. (*Note:* Do this carefully, in case you missed a spot and want to iron it some more.) You should now have the pattern transferred onto the copper side of the board. Immerse the board in a plastic tank filled with an etching solution for 10 to 20 minutes. I use a fish tank's filter reservoir. I also use a fish-tank air pump and some flexible plastic hose connected to about 8 inches of hard plastic pipe. The hard plastic pipe is then immersed in the reservoir. The bubbling oxygen from the pipe helps speed up the etching process.

Once the board is etched in acid and the exposed copper is removed, remove the toner with fine sandpaper and alcohol. Then, drill holes for the component leads. Harbor Freight Tools has a suitable drill press for $50. The drill bits are quite small and can be picked up at computer and ham flea markets for as little as $3 a set.

It is difficult to drill accurate holes for the IC sockets. A trick for doing this is to drill two holes as accurately as possible, and then fasten the board to a predrilled board. This can be done with a piece of 18- to 20-gauge copper wire through the first two holes. Then, use the holes in the predrilled board to align the holes you drill in the board you're making. I used this trick on the 68HC11 board and accidentally drilled a few extra holes. It is hard to tell where to drill the holes when drilling from the predrilled side. The extra holes didn't hurt anything; however, this procedure does seem to lead to more broken drill bits than usual.

Chapter 1

Digital Projects

In this chapter you will discover:

Power On Self-Test Card 1-1

A friend of mine once spent over $150 on a "power on self test" card. I advised him that they should cost around $50 instead. To make sure of my price, I designed and built a POST card in one evening. "Power On Self Test" or POST is the diagnostic every BIOS runs on the computer, when it is first turned on, to determine what is installed and what is working. As it runs each test, the BIOS outputs a code number to tell what test it is running. This test code number can be displayed with a "POST" card.

All that is needed to make a POST card is an address decoder to select address 80 (hex), an 8-bit latch, LED decoders, and LED displays. The address decoder consists of a 74LSO2 and a 74LS30. They are set up to look for address bit A7 high, the other address bits low, and I/O write low. Address bit A8 also needs to be low, and since I was short a low sensing input, I skipped A2. A2 can either be high or low, depending on the BIOS, so it is best to ignore it. Then a 74LS373 latches the 8 bits of data.

Professional devices use special decoders that correctly decode hexadecimal data. They also add several LEDs to monitor other things such as the power supplies and clocks. For testing power supplies I use a voltmeter, and for testing the clocks I use a logic probe.

I have added LEDs to monitor the status of memory and I/O read and write signals. This is to see if the motherboard is in fact running at all.

Pin	Signal	Pin	Signal
B13	I/O WRITE	A2	D7
A23	A8	A3	D6
A24	A7	A4	D5
A25	A6	A5	D4
A26	A5	A6	D3
A27	A4	A7	D2
A28	A3	A8	D1
A29	A2	A9	DO
A30	A1	B1	GND
A31	AO	B3	5 VDC

Table 1-1: IBM computer bus pinout

The IBM computer bus pinout is shown in Table 1-1. The "A" side is the top side of the card and the "B" side is the bottom. Pin 1 is toward the back of the computer and pin 31 is toward the front. The circuit boards usually have the pins marked on them.

A basic program can be used to run a simple test of the card. It will display the numbers in hexadecimal from "0" to "FF." The "FF" will be a double blank because the 7447s don't support hexadecimal. Table 1-2 shows some of the other incorrect translations.

Included is a sample BASIC program to test the POST card. Change the B=1 TO 1000 to B=1 TO 5000 for Pentium computers. It is a delay to make it easier to view the numbers as they are being displayed.

Hexadecimal	Appears as
"A"	looks like a small "c"
"B"	looks like a backwards small "c"
"C"	looks like a small "u"
"D"	looks like a small "c" with a line under it
"E"	looks like an upside-down "F"
"F"	comes up blank

Table 1-2: Incorrect translations from hexadecimal

01	CPU registers	0D	Reset test
02	CPU test done	0E	Refresh period
03	ROM (BIOS) checksum	10	64K base memory
04	8259 interrupt	11	Address line test
05	CMOS test	12	64K memory OK
06	VIDEO and timer	13	Interrupts started
07	CH 2 of 8253	14	Keyboard controller
08	CH 2 of 8254	15	CMOS R/W test
09	CH 1 of 8254	16	CMOS checksum
0A	CH 0 of 8254	17	Mono test
0B	Parity Check	18	Color test
0C	Refresh test	19	Video ROM test

Table 1-3: Some AMI POST codes

When the POST code reaches 20, the screen should come up and start displaying some information; otherwise the failure is either the graphics card or the ISA slot.

```
FOR A = 1 TO 256
  OUT 128, A        'TEST DATA TO OUTPUT
  FOR B = 1 TO 1000
  NEXT B            'DELAY FOR VIEWING
NEXT A
```

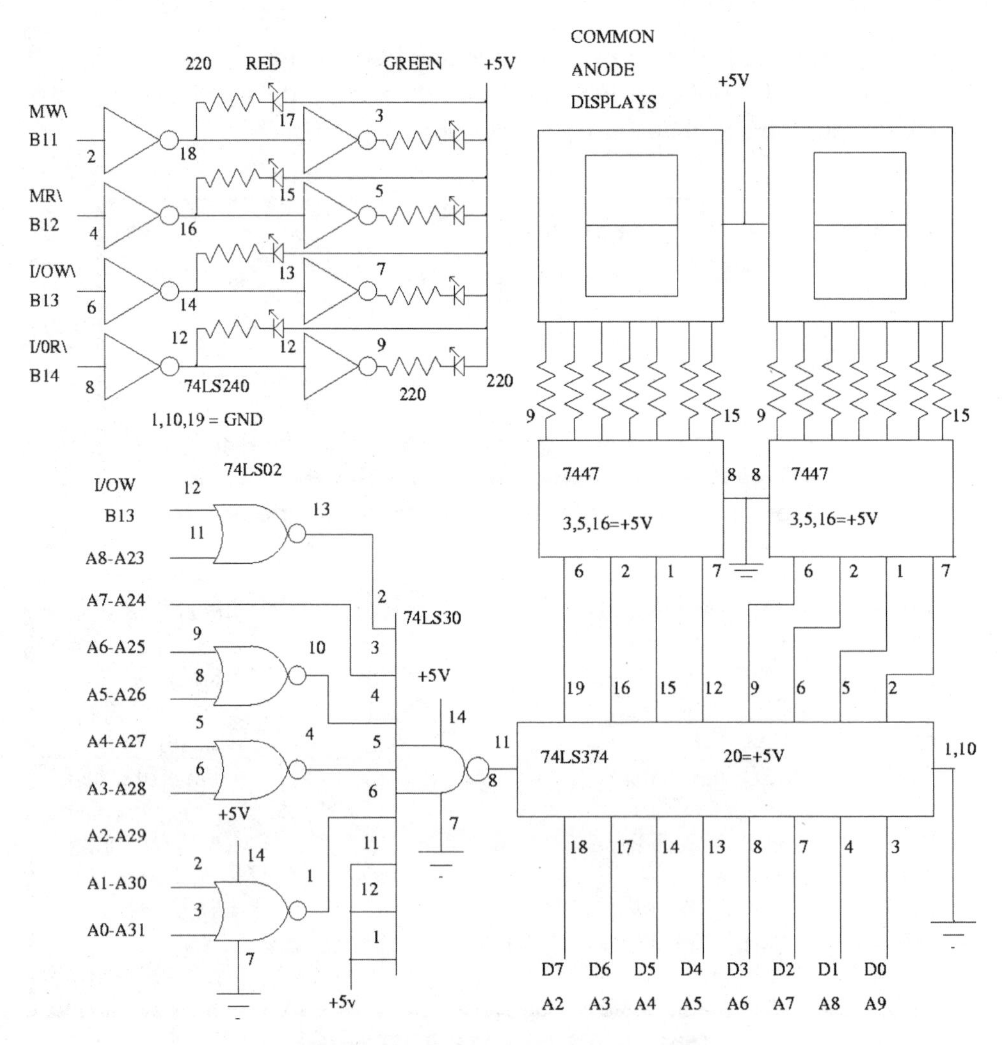

Figure 1-1: Power on self-test card

2	20-pin sockets
2	16-pin sockets
2	14-pin sockets
1	74LS374
1	74LS240
2	74LS47
1	74LS02
1	74LS30
2	Common anode LED displays (LA-6460 works fine)
20	220 ohm resistors (can use 330 ohms for less brightness)
1	Circuit board with PC XT edge connector pins

Table 1-4: Parts list

Figure 1-2: Picture of a POST card

Two-Chip BIOS to One Socket Adapter 1-2

A while back I purchased some 386 motherboards for $10 each. They were in various states of disrepair. With the aid of my POST card and landmark BIOS diagnostics, I was able to fix a couple of them.

Then there was a problem. The diagnostic BIOS was in two chips, and some 386 motherboards only take one. With some figuring, it was possible to make an adapter to allow the two-chip BIOS to run on the motherboards that have only one socket.

Address bit 0 had to be "anded" with the chip enable to select which chip was enabled. Then all of the address switches had to be shifted one position to correct for the missing Address bit 0. A 74F00 was selected to do the chip selection because it was one of the few fast chips on hand.

In Figures 1-3 and 1-4, you will see that I used a 28-pin socket under the header to space it up so that it will fit in tight spaces on some circuit boards. On some boards two or more of these extending sockets were needed.

The header for the 27512 socket has the small circuit board fastened to it. The board is just big enough to fit the two 28-pin sockets and the 14-pin socket, with one row of spare pads around the edge. The header's pins 15 to 28 are soldered to pads next to the same pins on one of the 28-pin sockets, in such a manner that the board ends up perpendicular to the header. This is done to connect the header to the circuit board.

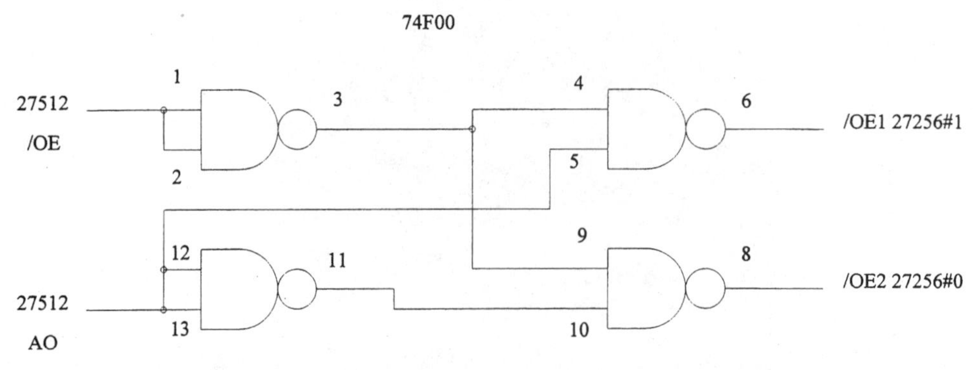

Figure 1-3: One-chip to two-chip select circuit

Signal Name	27512 Header	27256 Number 1	27256 Number 2
A15	1	27	27
A12	2	23	23
A7	3	4	4
A6	4	5	5
A5	5	6	6
A4	6	7	7
A3	7	8	8
A2	8	9	9
A1	9	10	10
A0	10 TO 74F00		
D0	11	11	11
D1	12	12	12
D2	13	13	13
GROUND	14	14	14
D3	15	15	15
D4	16	16	16
D5	17	17	17
D6	18	18	18
D7	19	19	19
CHIP ENABLE	20	20	20
A10	21	24	24
OUTPUT ENABLE	22 TO 74F00	FROM 74F00	FROM 74F00
A11	23	21	21
A9	24	25	25
A8	25	3	3
A13	26	2	2
A14	27	26	26
VCC + 5 V	28	1, 28	1, 28

Table 1-5: Pin/Signal translation table (to shift A0 one position)

1	28-pin header
2	28-pin sockets
1	14-pin socket
1	74F00
1	small "pad per hole" type circuit board.

Table 1-6: Required materials

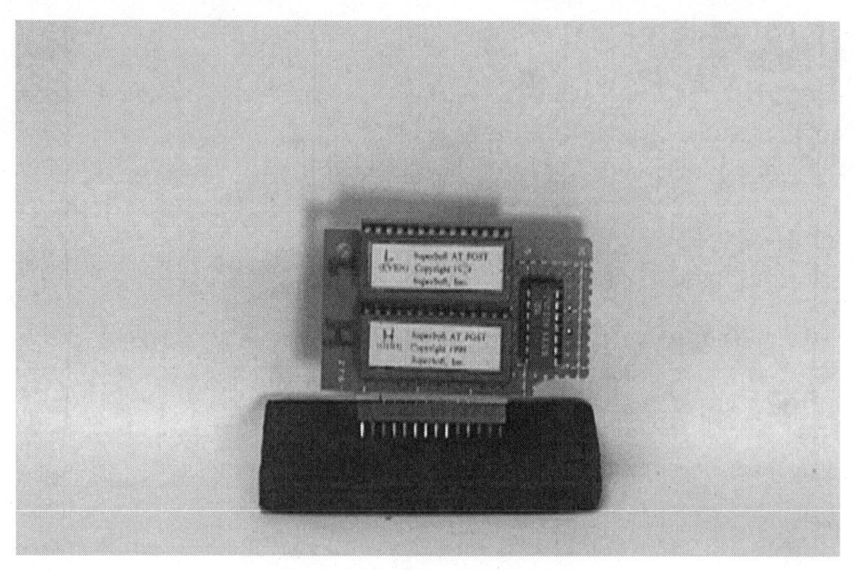

Figure 1-4: Picture of a two-chip BIOS to one socket adapter

Once the board is assembled and tested, they can also be glued together to make the connection stronger. The rest of the connections are made with 30-gauge wire-wrap-type wire. The wire is soldered point-to-point for all connections. You could use wire wrap, as the pins would be over the top of the header. It would then be difficult to solder the board fastening the pins to the header, and hard to reach the pins on the header.

IBM PC to CNC Adapter 1-3

When you shop for CNC (computer numerical control) equipment, one of the many overpriced devices you'll find is the floppy drive interface. It sometimes costs as much as $4,000 for a floppy drive. It is much easier to get an old PC laptop computer for $500 or less, make an interface cable, and set up some software. The PC allows you to edit the programs or break them down into smaller programs. There are plenty of prewritten off-the-shelf communication software packages available, too. Windows 3.1 terminal will work for this application.

There are problems with off-the-shelf software packages. Larger CNC programs won't fit into the CNC's memory for single stepping test purposes—unless, of course, you buy overpriced extra memory. When you're using available PC software, it won't tell you what it is going to send before it sends it, and the CNC executes it. So I developed this handy little program.

This is a scaled-down, send-only version of the software. It still allows you to see both the last ten lines sent (what the CNC is doing) and the next ten lines to be sent (what the CNC will be told to do next). No other off-the-shelf software comes close to having this feature.

In a large setup, a switch from the PC with four or more positions can determine which CNC machine it is controlling. Programs then can be uploaded to the CNC, or the CNC can single-step for fixing bugs. The XON/XOFF protocol is used as well as a baud rate of 4800, even parity, 7 bits, and 2 stop bits. This works fine with MORI-SEKI controllers. All of these parameters can easily be changed if desired.

A communications cable is easy to make. It is basically a null modem cable, with only send and receive connected. Table 1-7 shows the wiring needed for one that works quite well. The pins listed are for 25-pin connectors. Some newer computers may require a 9-pin connector. You can use a 9 to 25-pin adapter as needed.

The source listing is for Quick BASIC. The default communications rate is 4800, even parity, 7 bits, and 2 stop bits because most of our CNC machines used that rate. You can change the communications specifications in the software at the line that starts with "MCOMM$." You could also write a fancier version that prompts you for your communications specifications.

PIN	Signal	PIN
1 Not used		1 Not used
2	Data	3 (crossed)
3	Data	2 (crossed)
4 Jumper to 5		4 Jumper to 5
5 Jumper from 4		5 Jumper from 4
6 Jumper to 8 and 20		6 Jumper to 8 and 20
7	Ground	7
8 Jumper from 6		8 Jumper from 6
20 Jumper from 6		20 Jumper from 6

Table 1-7: Communications cable (only three wires run between the ends)

The CNC machine must use XON and XOFF to tell the program when it is able to receive more data. Pressing the Escape key will shut the program down. Always keep a hand near the CNC's STOP button as well, when troubleshooting new code.

```
'  PROGRAM PC_NC (IBM PC TO NUMERICAL CONTROL INTERFACE)
'  WRITTEN 2-19-93 BY BOB DAVIS
COLOR 7, 1: CLS
DIM LINE$(12): DIM OLDLINE$(12)
FOR VPOS = 8 TO 13: LOCATE VPOS, 61: PRINT "|": NEXT VPOS
LOCATE 7, 62: PRINT "__________________"
LOCATE 10, 63: PRINT "SEND? (Y/Q)"
LOCATE 11, 63: PRINT "COMM. SPECS.;"
LOCATE 12, 63: PRINT "4800, E, 7, 2"
LOCATE 13, 62: PRINT "__________________"
GETFILE: LOCATE 9, 63: INPUT "FILE: ", NCFILE$ 'GET FILE
IF NCFILE$ = "" THEN CLS : SYSTEM
OPEN NCFILE$ FOR INPUT AS #1
MREADY$ = "N"
LOCATE 10, 74: INPUT MREADY$
IF MREADY$ <> "Y" AND MREADY$ <> "y" THEN SYSTEM
MAINLOOP:
MCOMM$ = "COM1:4800, E, 7, 2, CD1000"
CLOSE #2: OPEN MCOMM$ FOR RANDOM AS #2
VPOS = 10: LASTLINE$ = "F"
CLEARLINE$ = "                                                          "
DO WHILE LASTLINE$ = "F"
   IF EOF(1) THEN LINE$(11) = " " ELSE INPUT #1, LINE$(11)
   IF LINE$(0) = " " THEN
      LASTLINE$ = "T"
   END IF
   VPOS = 2: VPOS2 = 13: COLOR 10, 1
   OLDLINE$(11) = LINE$(0)
   FOR K = 0 TO 10          ' DISPLAY LINES TO SEND
      LOCATE VPOS2 + K, 3: PRINT CLEARLINE$
      LOCATE VPOS2 + K, 3: PRINT LEFT$(LINE$(K), 58)
      COLOR 7, 1            ' RESET COLOR TO NORMAL
      LOCATE VPOS + K, 3: PRINT CLEARLINE$
      LOCATE VPOS + K, 3: PRINT LEFT$(OLDLINE$(K), 58)
   NEXT K
   FOR L = 1 TO LEN(LINE$(0))
      PRINT #2, MID$(LINE$(0), L, 1);
      IF NOT EOF(2) THEN XCHECK$ = INPUT$(1, #2)
      IF XCHECK$ = CHR$(19) THEN      ' CHECK FOR XON AND XOFF
         DO WHILE XCHECK$ <> CHR$(17)
            IF INKEY$ = CHR$(27) THEN GOTO MAINLOOP
            IF NOT EOF(2) THEN XCHECK$ = INPUT$(1, #2)
         LOOP
      END IF
```

```
    NEXT L
    PRINT #2, CHR$(10)                    ' SEND LINE FEED
    FOR K = 0 TO 10                       ' SCROLL UP A LINE
        OLDLINE$(K) = OLDLINE$(K + 1)
        LINE$(K) = LINE$(K + 1)
    NEXT K
    IF INKEY$ = CHR$(27) THEN GOTO MAINLOOP ' ESCAPE PRESSED
LOOP
CLS : SYSTEM
```

Barcode Printing Program 1-4

This project primarily describes software. You can't easily make the hardware yourself—you have to buy it. The problem is that you can purchase bar-code reading devices for as little as $100. Then, you will find that the software that comes with it can do very little. They want you to send them more money for a more powerful version. So I developed some software to do the job in Quick BASIC. At first, I used the block characters with half a solid character for a thin line and one and a half characters for a wide line. This worked amazingly well, but if the bar code was more than seven digits long, it took up the entire width of the paper! You also had to scan the code very fast to "fake out" the wand.

Bar codes aren't that difficult to develop, I've never had formal training on bar codes, and don't know the "right" way to make them. So I studied some bar codes and figured out how they work. Each character converted into a bar code is represented as one wide space, two wide bars, and three narrower bars. These are represented in my program as the letters "S" for space, "W" for wide, and "N" for narrow. Each of the 39 possibilities has a translation to the needed bar-code combination.

The real trick is the graphics portion. Using characters was easy, but printing graphic bytes is something else. This program was written for an Okidata 24-pin printer, but should work on others with little modification. It has been tested on several laser printers with no problems.

Each bar is 6 bytes wide, but on a 24-pin printer, that requires 3 bytes to be sent for each of the 6 bytes printed, for a total of 18 bytes. A narrow line is then 18 lines and 18 spaces. A space then is 36 spaces wide. A wide line is 54 bytes and 18 spaces. Each character is then 1 space (36 bytes), 2 wide lines (144 bytes), and 3 narrow lines (108 bytes), for a total of 288 bytes per character. This is then divided by 3 for the graphics code to be sent to the printer, for a result of 96 bytes per character.

This complicated result is then plugged into a formula, where it is multiplied by the number of characters to send, and then 2 is added for start and stop codes. This is divided by 256 to get the low number. The remainder from that division is then also used as the high number. These are then added to a string of control codes and output to the printer to tell it how many graphic bytes will be coming its way. If the numbers are wrong, the printer will try to print the codes and respond in very strange ways.

Basically, you have to tell the printer that the following "X" number of bytes will be graphics data. If the math is wrong, the printer will see data as things like line feeds and carriage returns.

Once again, this is a scaled-down version of the program with a lot of features removed. But, as you can see, you can add features to your heart's desire, as well as position the BARCODE anywhere you might want it. With a little software conversion, this program works in Clipper or Dbase as well.

```
' PROGRAM: BAR-CODE.BAS, BY: BOB DAVIS, WRITTEN: 01/20/95
CLS
PRINT "BARCODE PRINT PROGRAM"
INPUT "BARCODE: ", MNAME$
WIDTH "LPT1:", 255
LENGTH = LEN(MNAME$) + 2              ' 1 start + length + 1 stop
LOWNUM = INT(LENGTH * 96 / 256)     ' 4 LINES * 3 BYTES * 8 MCODES
HINUM = 96 * LENGTH - LOWNUM * 256' REMAINDER FROM LOWNUM DIVISION
BARC$ = CHR$(27) + "*(" + CHR$(HINUM) + CHR$(LOWNUM)
FOR BYTE = 1 TO LENGTH
IF BYTE > 1 THEN mletter$ = MID$(MNAME$, BYTE - 1, 1) 'LETTER
IF BYTE = 1 THEN mletter$ = " ": mcode$ = "NSNWWN"     'STOP CODE
IF BYTE = LENGTH THEN mletter$ = " ": mcode$ = "NSNWWN" 'START
IF mletter$ = "A" THEN mcode$ = "WNNSNW"
IF mletter$ = "B" THEN mcode$ = "NWNSNW"
IF mletter$ = "C" THEN mcode$ = "WWNSNN"
IF mletter$ = "D" THEN mcode$ = "NNWSNW"
IF mletter$ = "E" THEN mcode$ = "WNWSNN"
IF mletter$ = "F" THEN mcode$ = "NWWSNN"
IF mletter$ = "G" THEN mcode$ = "NNNSWW"
IF mletter$ = "H" THEN mcode$ = "WNNSWN"
IF mletter$ = "I" THEN mcode$ = "NWNSWN"
IF mletter$ = "J" THEN mcode$ = "NNWSWN"
IF mletter$ = "K" THEN mcode$ = "WNNNSW"
IF mletter$ = "L" THEN mcode$ = "NWNNSW"
```

```
IF mletter$ = "M" THEN mcode$ = "WWNNSN"
IF mletter$ = "N" THEN mcode$ = "NNWNSW"
IF mletter$ = "O" THEN mcode$ = "WNWNSN"
IF mletter$ = "P" THEN mcode$ = "NWWNSN"
IF mletter$ = "Q" THEN mcode$ = "NNNWSW"
IF mletter$ = "R" THEN mcode$ = "WNNWSN"
IF mletter$ = "S" THEN mcode$ = "NWNWSN"
IF mletter$ = "T" THEN mcode$ = "NNWWSN"
IF mletter$ = "U" THEN mcode$ = "WSNNNW"
IF mletter$ = "V" THEN mcode$ = "NSWNNW"
IF mletter$ = "W" THEN mcode$ = "WSWNNN"
IF mletter$ = "X" THEN mcode$ = "NSNWNW"
IF mletter$ = "Y" THEN mcode$ = "WSNWNN"
IF mletter$ = "Z" THEN mcode$ = "NSWWNN"
IF mletter$ = "-" THEN mcode$ = "NSNNWW"
IF mletter$ = "." THEN mcode$ = "WSNNWN"
IF mletter$ = "1" THEN mcode$ = "WNSNNW"
IF mletter$ = "2" THEN mcode$ = "NWSNNW"
IF mletter$ = "3" THEN mcode$ = "WWSNNN"
IF mletter$ = "4" THEN mcode$ = "NNSWNW"
IF mletter$ = "5" THEN mcode$ = "WNSWNN"
IF mletter$ = "6" THEN mcode$ = "NWSWNN"
IF mletter$ = "7" THEN mcode$ = "NNSNWW"
IF mletter$ = "8" THEN mcode$ = "WNSNWN"
IF mletter$ = "9" THEN mcode$ = "NWSNWN"
IF mletter$ = "0" THEN mcode$ = "NNSWWN"
FOR MCPOS = 1 TO 6
   IF MID$(mcode$, MCPOS, 1) = "S" THEN
      FOR num = 1 TO 36
         BARC$ = BARC$ + CHR$(0)
      NEXT num
   END IF
   IF MID$(mcode$, MCPOS, 1) = "N" THEN
      FOR num = 1 TO 18
         BARC$ = BARC$ + CHR$(255)
      NEXT num
      FOR num = 1 TO 18
         BARC$ = BARC$ + CHR$(0)
      NEXT num
   END IF
   IF MID$(mcode$, MCPOS, 1) = "W" THEN
      FOR num = 1 TO 54
         BARC$ = BARC$ + CHR$(255)
      NEXT num
      FOR num = 1 TO 18
         BARC$ = BARC$ + CHR$(0)
      NEXT num
   END IF
NEXT MCPOS
NEXT BYTE
LPRINT CHR$(27); CHR$(48);                          '1/8" SPACING
LPRINT CHR$(15); CHR$(27); CHR$(77);                '20 CPI
```

```
LPRINT BARC$: LPRINT BARC$: LPRINT BARC$
LPRINT CHR$(27); CHR$(50);                         '1/6" SPACING
LPRINT CHR$(18); CHR$(27); CHR$(80);               '10 CPI
LPRINT "": LPRINT MNAME$
LPRINT "": LPRINT ""
```

EPROM Copiers 1-5

Most of the time you don't really need an EPROM programmer to program an EPROM. All you really need is a method of copying everything from an existing EPROM to another one. So I developed a simple three-IC circuit that can be built for as little as $20. This compares to spending more than $250 for even the cheapest of EPROM programmers. Over the years my copier has expanded to cover larger EPROMs as computer motherboards, and various projects, have used larger ones for their BIOSs.

An EPROM is an "electronically programmable read-only memory." However, despite their name, they can be written to—once. The most common type can then be erased with an ultraviolet light. This can be done by placing it in direct sunlight for a day, or you can use a UV eraser available for $30 to $50.

I personally use a homemade ultraviolet eraser. It was removed from an ultraviolet light assembly inside a dehumidifier. Some dehumidifiers use UV light to keep "stuff" from growing in their water storage trays. It works quite well and erases an EPROM one-quarter inch from the bulb in about 1 hour.

Once they're programmed, EPROMs will need a cover to protect them from being erased by stray UV light. A write-protect tab from a 5-1/4 inch floppy works great as a cover.

Sometimes the needed existing EPROM doesn't exist. The source EPROM can be emulated by copying the data into battery-backed SRAM. Then it can be copied into an EPROM. In some cases I've used a breadboard to build the logic to develop the code necessary to program an EPROM.

The most common EPROM is the BIOS found on motherboards in every computer. The BIOS on a motherboard is what tests the computer, and starts it up. BIOS stands for "basic input output system." It is encoded with software that first tests the computer to determine what kind of devices are installed and see if they are working properly. Then, once it has determined what the resources are, it tries to get them to start operating.

In the early days, the BIOS had to be replaced in order to add new features. Today it can be reprogrammed with software. If the BIOS needs to be updated, you can download the needed software and reprogram it. But if it becomes bad, you don't need a new motherboard—just replace the BIOS with a reprogrammed one. You can make a copy of a good BIOS by determining the type of EPROM it is, finding a good one on an identical motherboard, and copying it. Don't forget to transfer the license sticker to the new chip.

EPROM Copiers: 2764 1-6

The oldest EPROM BIOS covered here was once used in PCs and XTs. They had allotted 8K addresses by eight bits for the BIOS. That translates into a 64K BIOS EPROM chip. Those computers also had several extra sockets for more EPROMs to hold BASIC. These sockets were eventually dropped from motherboards because no one ever used them. Some of these old motherboards even required a ROM-to-EPROM adapter as the pinout was different.

The 2764 is supposed to have 50-ms programming pulses. To get them, I used 60 hertz (i.e., the power line) for a 16-ms clock, then used a circuit to develop a three-of-eight duty cycle. Three times 16 is 48 ms, which is well within the tolerance of the 2764. The programmer then cycles through all of the addresses. At each address, it repeatedly copies the data from the source EPROM to the destination EPROM.

My first application for the copier was to replace two ROMs and some discrete logic with one EPROM. To do this, I had to copy two 2513 uppercase and lowercase character generators into a 2764. Then I added block graphics. That was done with a breadboard source EPROM emulator. The resulting chip was used in my homemade TRS-80 clone to generate characters.

74LS393 pin 3	0 1 0 1 0 1 0 1 0 1
74LS393 pin 4	0 0 1 1 0 0 1 1 0 0
74LS393 pin 5	0 0 0 0 1 1 1 1 0 0
74LS02 pin 4	1 0 0 0 1 0 0 0 1 0
74LS02 pin 10	0 1 1 1 0 0 0 0 0 1
74LS02 pin 13	1 0 0 0 1 1 1 1 1 0

Table 1-8: Timing chart

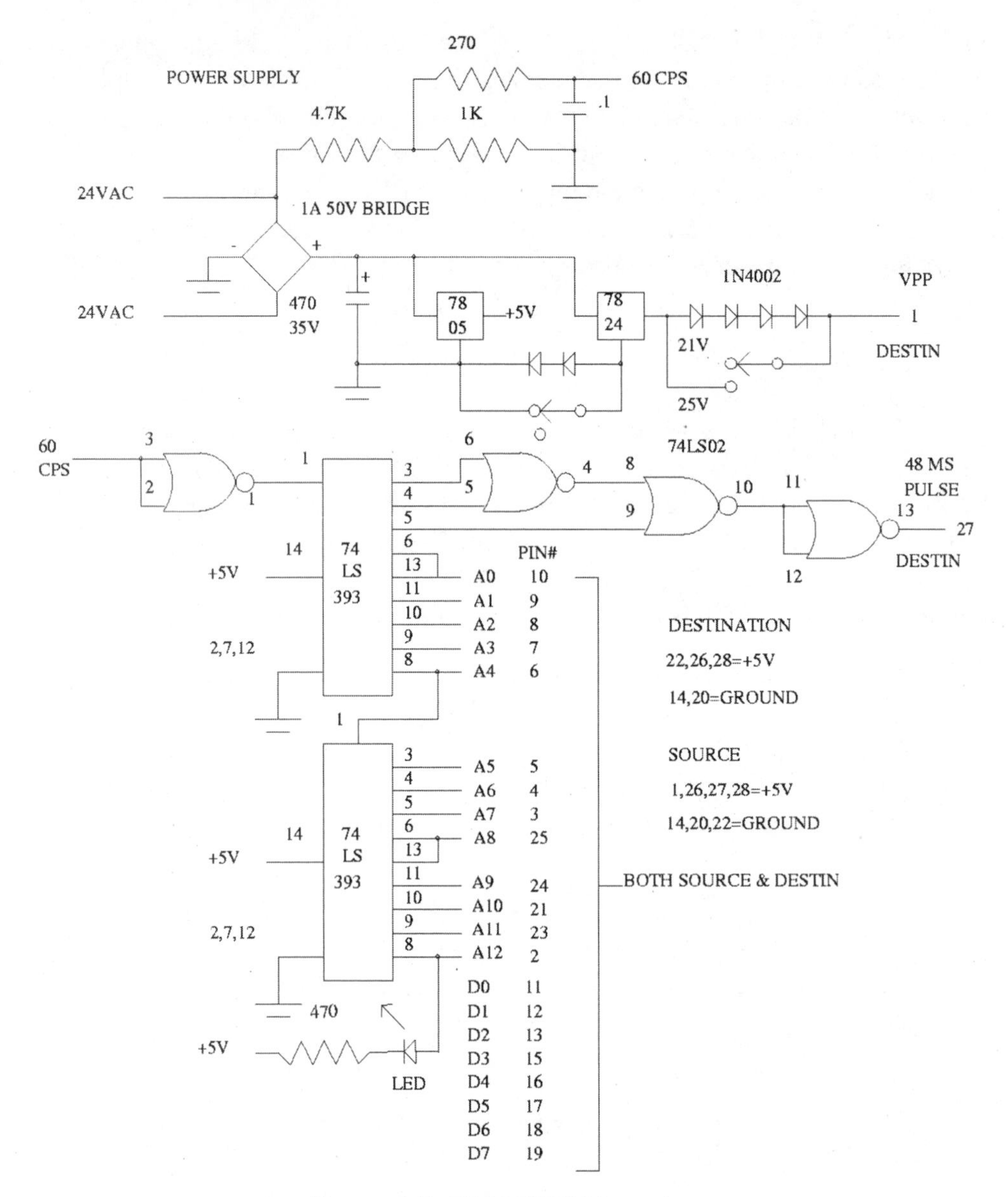

Figure 1-5: 2764 EPROM copier

1	28-pin socket (can be a ZIF socket)
3	14-pin sockets
1	74LS02
2	74LS393
1	24VAC adapter
1	50-volt, 1-amp bridge rectifier
1	LM7805 voltage regulator
1	LM7824 voltage regulator
6	1N4002 rectifier diodes
1	DPDT switch
1	circuit board, about 3 × 6 inches
Miscellaneous	resistors and filter caps
Optional	8 LEDs to see the current address

Table 1-9: Parts list

Later this copier helped me repair corrupted BIOSs on some XTs. It was also used with a 2732 adapter to make dual EPROMs for Eagle motherboards. This made them 100% IBM XT compatible. It worked like the 2-to-1 BIOS adapter, but in reverse.

The power supply provides a 4-volt, 60-hertz reference for the timing, as well as several voltages. Diodes are used with the 7824 regulator to adjust it up 2 × 0.6 volt, or 1.2 volts higher. With the switch, those diodes are shorted out, and four diodes are brought in series with the regulator output to reduce the 24 volts by 4 × 0.6 volts, or 2.4 volts total. This then provides 21.6 volts for EPROMs that require the lower voltage. Thus, you can select either 25.2 or 21.6 volts, depending on the requirements of the EPROM to be programmed.

EPROM Copiers: 27256 1-7

After the XT computer, the next computer was the AT computer. With the AT came the need to repair corrupted 27256 BIOSs. This need was first clearly demonstrated when an unreliable computer failed every time you turned on the lights! The BIOS was "borderline" and the light was almost erasing a bit or two from memory.

In an AT computer, two chips are used at 32K addresses by 8 bits each (256K EPROM) for a total of 32K by 16 bits. But you thought only 8K was allotted for the BIOS? I think the designers of the AT computer decided to use the unused space that had been allotted for BASIC in ROM to give more room for a larger BIOS. This additional space was allotted at first for hard drive setup tables, but eventually came to include a complete setup stored in EPROM.

This allotment of memory can be seen by using my printer port EPROM verifier on some older Pentium BIOSs. At first, they only used space in the BIOS from 40K to 64K, and the rest of the BIOS was empty. That amounted to three blocks of 8K that was actually used. Later, the entire lower 64K appears to have been used.

I liked the earlier copier design so much that I improved it for the larger chips. This time it had to generate 1-ms programming pulses repeatedly for each address. A 555 timer was used at four times the desired pulse width. A new 74LS393 was added to divide it by 4, and to set the amount of programming pulses per address. I settled on 16 programming pulses per address, but that can easily be changed.

The new 74LS393 was also used to make sure that the address didn't change just before or during a programming pulse. This is done by inverting the "QB" output with a 2N2222 and using it to clock the second half of the 74LS393. Since the 74LS393 normally clocks on a high-to-low transition, it now clocks on

1	12-volt AC adapter
1	50-volt, 1-amp bridge
1	LM7805 voltage regulator
4	1N4002 diodes
1	DPDT switch
1	28-pin IC socket (can be a ZIF socket)
3	14-pin IC sockets
3	74LS393
1	8-pin socket
1	555 timer
1	2N2222 transistor
1	circuit board, about 3 × 6 inches
Miscellaneous	resistors and filter caps
Optional	8 LEDs and resistors to see the current address

Table 1-10: Parts list

the opposite. The programming pulse is active when low, so the address won't change until after the programming pulse, instead of just before it.

This time the power supply needed to provide 6 volts to the EPROM being programmed. This would tell it to enter the programming mode. Then, the other ICs

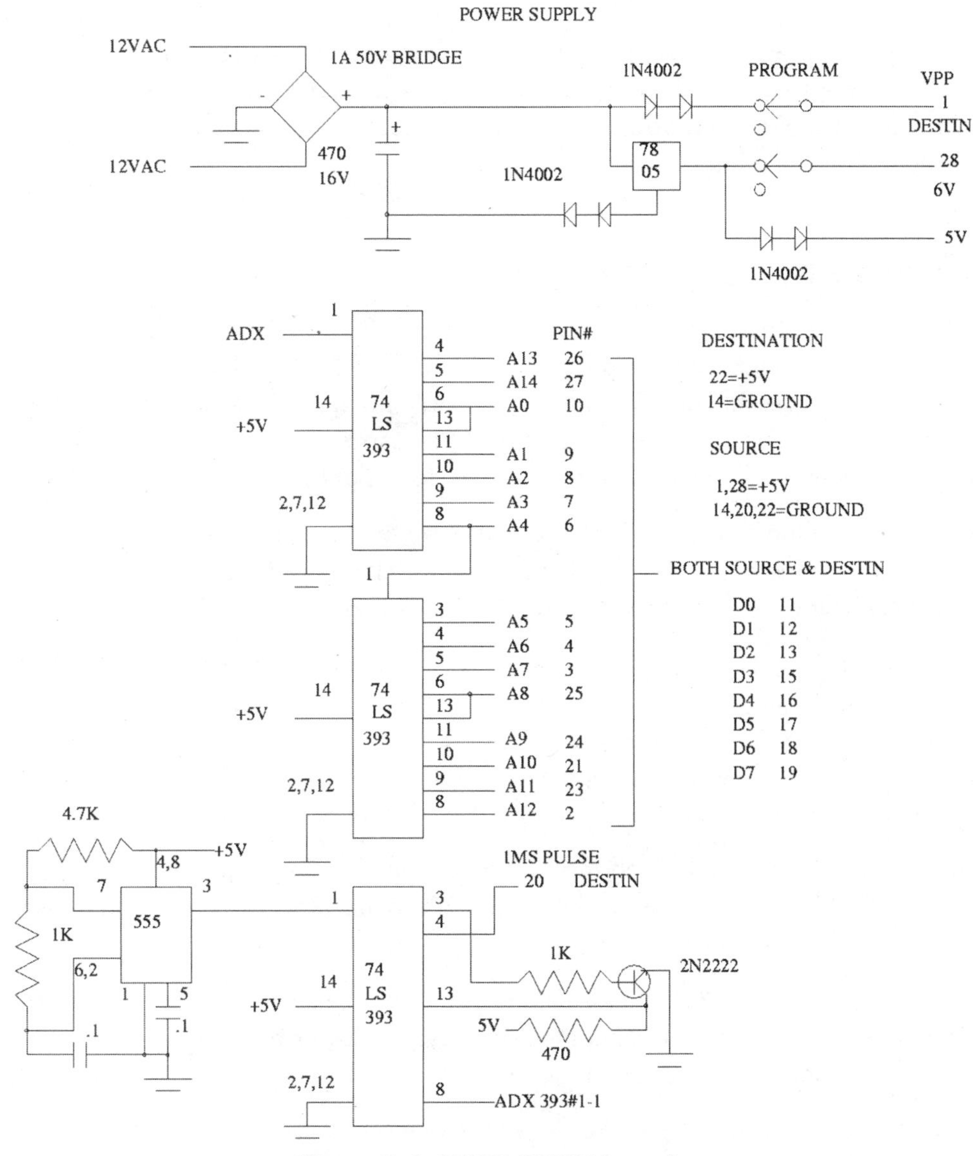

Figure 1-6: 27256 EPROM copier

need a normal 5-volt source. This was done with diodes to raise the 5 volts to 6.2 volts, and then more diodes reduce it back down to 5 volts again. Next, a source of 12.5 volts was needed, but the AC adapter in use didn't provide enough voltage to operate a regulator reliably. Once again, diodes were used to provide the voltage needed, but this arrangement varies with AC adapters, as well as the type of 27256 that is being programmed. The solution is to add extra diodes, meter the results, and remove any unnecessary diodes with jumper wires.

Not shown in the schematic, but needed, are filter capacitors of 0.01 to 1 μF per IC. You can also optionally add LEDs in series with 470-ohm resistors connected to 5 volts for the upper 8 address bits. This will let you know when it has cycled through all of the addresses.

EPROM Copiers: 27010 1-8

Pentium motherboards use a 1-megabyte BIOS chip. Usually it is a 28F1000, which is a 128K by 8-bit flash-programmable EEPROM. EEPROM stands for "electronically erasable programmable read-only memory. These do not need an ultraviolet light source for erasure. Instead, they can be erased electronically and then reprogrammed. Quite often this can be done right on the motherboard.

Two different pinouts are used, differentiated by the part number's suffix. If the part number ends in 1000 or 1001, then output enable is on pin 2 and A16 is on pin 24. If the part number ends in 010, those two pins are reversed.

The strange thing is that the part number doesn't matter on motherboards. Most likely they are using just 64K of the BIOS because that is all the memory space allotted for the BIOS. In other words, if you copy a 28F1000 into a 27C010, it will work just fine. I've done it repeatedly, but this may not be true for the newest motherboards.

The only problem would be on motherboards that have a jumper to select two different BIOSs within the same chip. If that were the case, then you would lose that ability in a copied BIOS.

The portion of the EPROM that is actually used can be seen with the printer port EPROM verifier. You will see that they only use the lower block of 64K. Some older BIOSs don't even use all of the 64K.

1	24-volt center tapped AC adapter (you may have to use a transformer)
1	50-volt, 1-amp bridge
1	LM7812 voltage regulator
1	LM7805 voltage regulator
5	IN4002 diodes
1	DPDT switch
1	28-pin IC socket (can be a ZIF socket)
3	14-pin IC sockets
3	74LS393
1	8-pin socket
1	555 timer
1	2N2222 transistor
1	circuit board, about 3 × 6 inches
Miscellaneous	resistors and filter caps
Optional	8 LEDs to see the current address

Table 1-11: Parts list

The prefix on an EPROM part number also tells us something. If the part number starts with 27, it is usually a UV-erasable EPROM. If it starts with 28, it is an electronically erasable model. There are other prefixes such as 23, which means the chip is a one-time programmable version or that it was mask-programmed at the factory.

How did we get back to only one BIOS chip? The AT computer used two chips because it was a 16-bit device, so the 386 or higher would then require four chips because they are 32-bit devices. Instead, the motherboard copies the BIOS into much faster 32-bit RAM, 8 bits at a time. It then runs the BIOS out of the faster 32-bit RAM. That way only one BIOS chip is needed. To the best of my knowledge this is true of most 486s and all Pentium motherboards.

If you need to copy two diagnostic chips into one, use the two chip BIOS to one socket adapter for the source EPROM. However, the results may not work—older diagnostic BIOSs have problems with newer Pentium motherboards. This is because the memory and cache configuration is proprietary to the motherboard's chip set.

The EEPROM offers the advantage over EPROMs of on-site reprogrammability. However, that ability has also caused problems. Motherboards have been shipped with the jumper on to allow easy programming of the BIOS. Programs

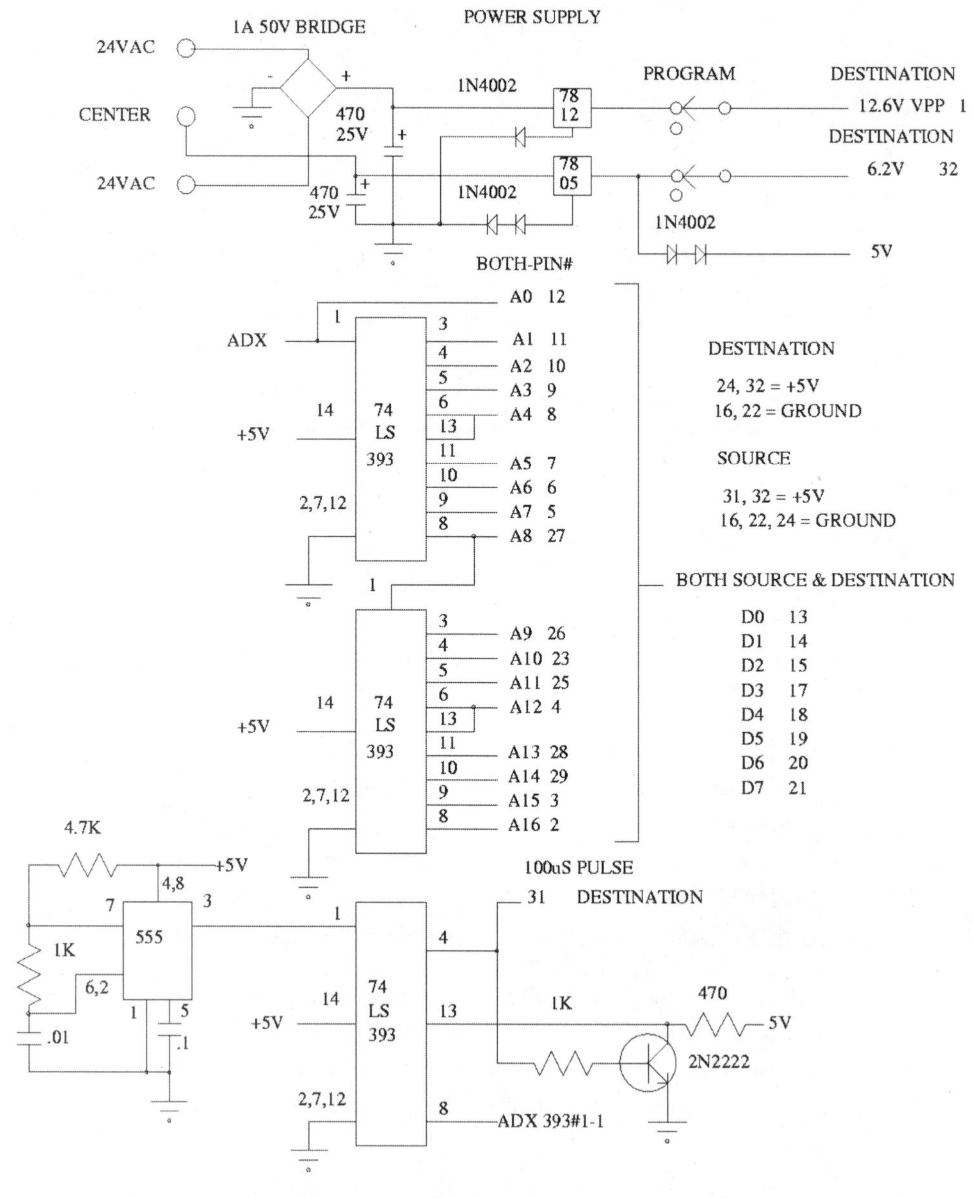

Figure 1-7: 27C010 EPROM copier

can then crash and overwrite the BIOS. They are also subject to power losses during reprogramming, which would destroy the BIOS program. It is even possible that a virus could overwrite the BIOS. There is, of course, operator error (i.e., copying the wrong BIOS for a particular motherboard!).

To fix a bad BIOS, first you have to find an identical working motherboard. Many times a motherboard with the same chip set will not work. The designers seem to change things in the cache configuration and in the built-in I/O. These changes will then confuse the BIOS. Once you find an exact match, borrow the BIOS from the good motherboard and copy it.

This time the programming pulse needs to only be 100 microseconds in length. This shorter pulse was obtained by changing the 0.1-μF cap on pins 6 and 2 of the 555 timer to 0.01 μF. I totally rebuilt the copier this time, so that the address pins were in the correct order.

The new power supply was designed to provide a well-regulated 12.6-volt source. However, some EPROMs may require a higher voltage, closer to 13

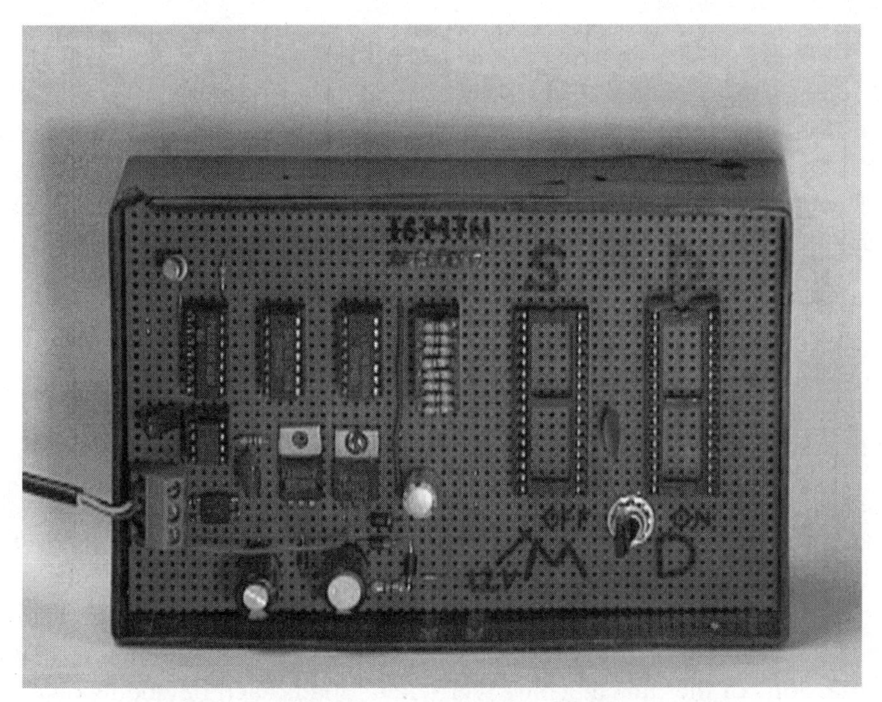

Figure 1-8: Picture of a 27C010 EPROM copier

volts. If this is the case, add a 33- or 47-ohm resistor in series with the diode on the ground pin of the 7812 regulator. By using a 24-volt center tapped transformer, I was able to obtain both 14 and 28 volts DC for the regulators.

Once again, I did not show the filter caps on the 12.6-, 6.2-, and 5-volt sources. A 0.1-μF, 25-volt cap on each will do. An 18-volt center tapped transformer might have worked better, as the regulators would not get as hot.

Quiz Machines 1-9

Older quiz machines used several ICs, or even relays. Some had 110-volt relays. With 110 volts going to the hand units, a potential shock hazard existed. The relay design required 2 relays per player or 16 relays for 8 players. This could also prove very expensive. Designs with several ICs quite often needed a delay so that the data would get captured. This delay can lead to ties, with two lights on. My design uses only one or two ICs and sends only a much safer level of 5 volts to the hand units.

I came up with this design way back in 1983. I was working with what was then new 8-bit latches. I wanted to see if the chip count could be reduced from designs that used the older 2- and 4-bit latches. With the 74LS374 8-bit latch, a delay was needed to keep it from coming up blank. But if the delay was too long, it would allow ties. I knew there must be a better way. I then tested the 74LS373, and it worked perfectly every time.

Instead of looking for an input and then latching the data (sometimes missing it because of noise), my new quiz machine design uses a "transparent latch" and looks at its own output. It is called "transparent" because, when it is not latched, it merely buffers data from its input to its output. That way, it cannot miss the data.

Originally, the first 74LS373 based quiz machine had an eight input gate that checked for data and enabled the latch. Then, in a newer version, the output indicators themselves were used to check for data. This method required a two-transistor amplifier to toggle the gate for a TTL level device. Later, in the battery-powered version, the LEDs themselves drove the gate directly. Then, other versions of the quiz machine were developed, each having its own advantages.

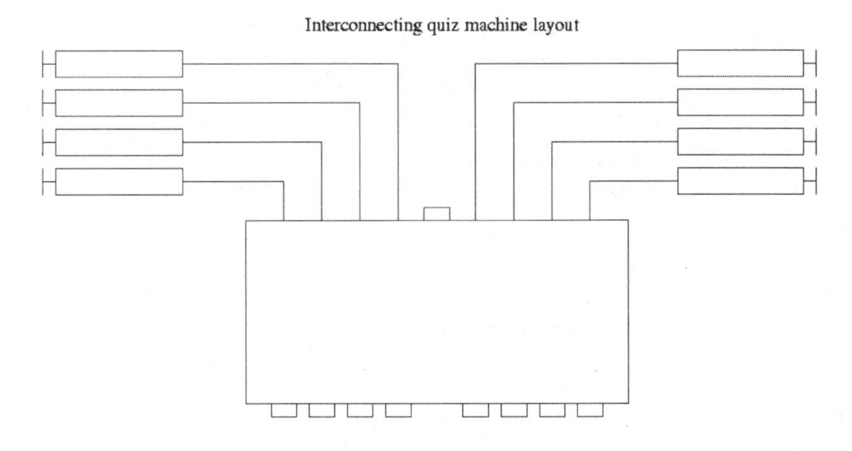

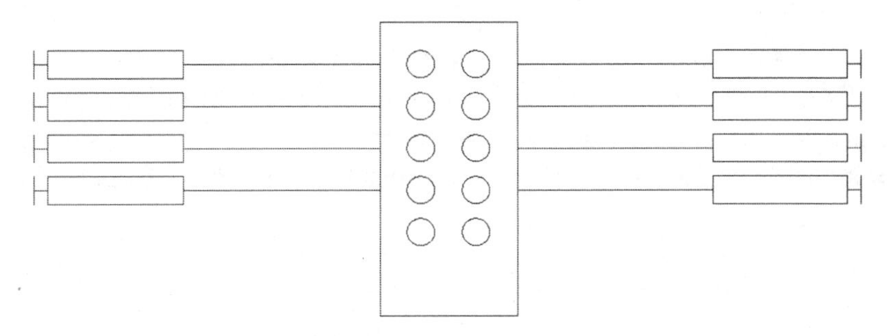

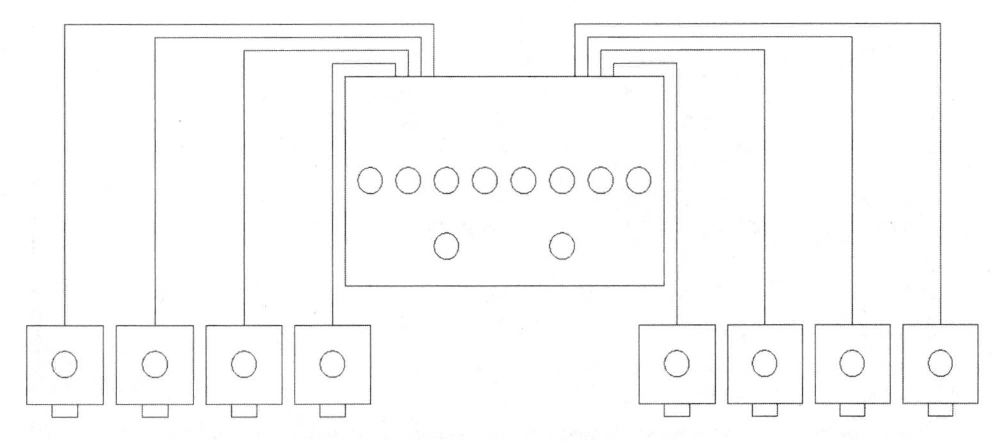

Figure 1-9: Quiz machine layouts

Quiz Machines: Interconnecting 1-10

The original version can be interconnected. This allows several teams to compete against each other at the same time. All the quiz machines had to share a 6-volt AC adapter (or 5-volt regulated supply) and a reset switch. The resets are hence "OR" tied together so one quiz machine can latch all other connected machines.

The reset switch forces the latch high, unlatching the latched machines. A 2N2222 transistor on the low end of the input switches sounds a buzzer whenever any switch is activated.

The box used is about 6 inches wide by 4 inches deep by 2 inches tall. The interconnect input can be a "stereo headphone" type jack. They also work independently with AC adapters and individual reset switches.

The layout I use has the lights evenly spaced along one of the 6-inch by 2-inch sides. The player's switches then go to the opposite side through eight holes. These are lined up with the lights. This way it is easy to see which switch belongs with what light. The top and bottom are unused so they can be stacked on top of each other when interconnected.

1	6 Volt DC-AC adapter (use a diode or 5-volt regulator to get 5 volts)
9	momentary switches
1	1/4 inch stereo headphone type jack
1	20-pin IC socket
1	74LS373
3	2N2222 transistors
1	box, 2 × 6 × 4 inches (approx.)
1	circuit board, about 1 × 2 inches
1	buzzer
2 feet	1/2-inch PVC pipe
8	LEDs and LED holders
32 feet	2 conductor stranded speaker type wire (4 feet per player)
Miscellaneous	resistors and filter caps

Table 1-12: Parts list

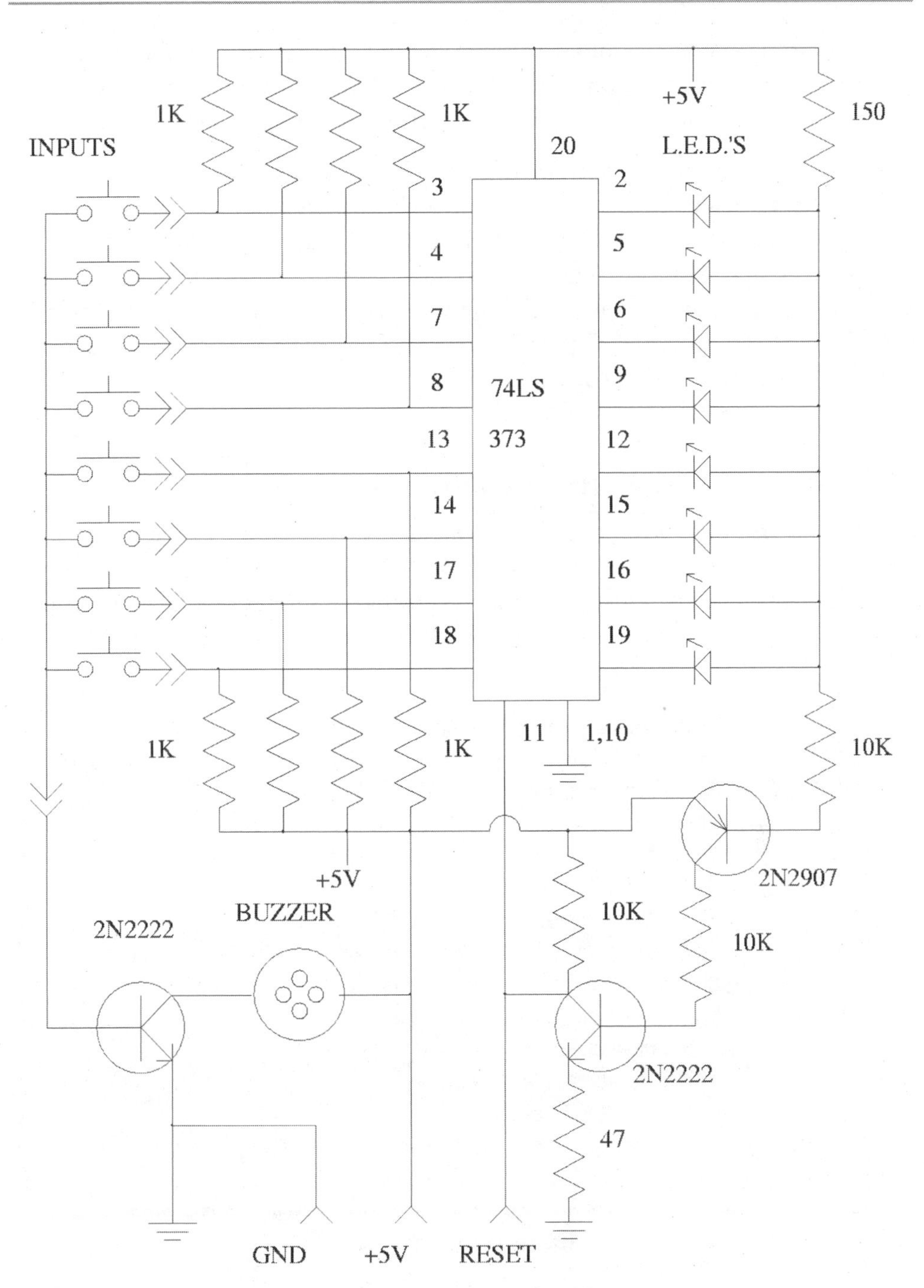

Figure 1-10: Interconnecting quiz machine

The reset and power plug into a 1/4-inch stereo jack like the one used for headphones. This jack is found on the back of the cabinet. To use several machines a special cable needs to be made with lots of plugs that interconnect the machines.

The hand units can be made from 1/2-inch PVC pipe. Cut the pipe into about 4-inch lengths. Then remove the nuts and lock washers from some Radio Shack 275-609 switches. Solder the wires to the switches, put glue on the threads of the switches, and slide the PVC piece up over the switch body. (*Note:* No heat-shrink tubing is needed.) Then, also glue the wire at the point where it comes out of the bottom of the PVC pipe lengths. This will keep the wires from accidentally being pulled off the switch.

Quiz Machines: Battery-Powered 1-11

The battery-operated version is truly portable. It was designed for practicing in the car and on the go. A "C" or "HC" series IC must be used for correct latching. The CMOS threshold for a logic zero is higher than that of TTL logic. The

1	9-volt battery
8	1N914 diodes (or equivalent)
9	momentary switches
1	SPST power switch
1	20-pin IC socket
1	74HC373
1	1N914
1	2N2222 transistor
1	box, 0.75 × 3 × 1.5 inches (approximate size of a calculator)
1	circuit board, about 1 × 2 inches
1	buzzer (small)
4	BIC pen shells
4	pieces heat-shrink tubing
8	LEDs and LED holders
32 feet	two-conductor stranded speaker type wire (4 feet per player)
Miscellaneous	resistors and filter caps

Table 1-13: Parts list

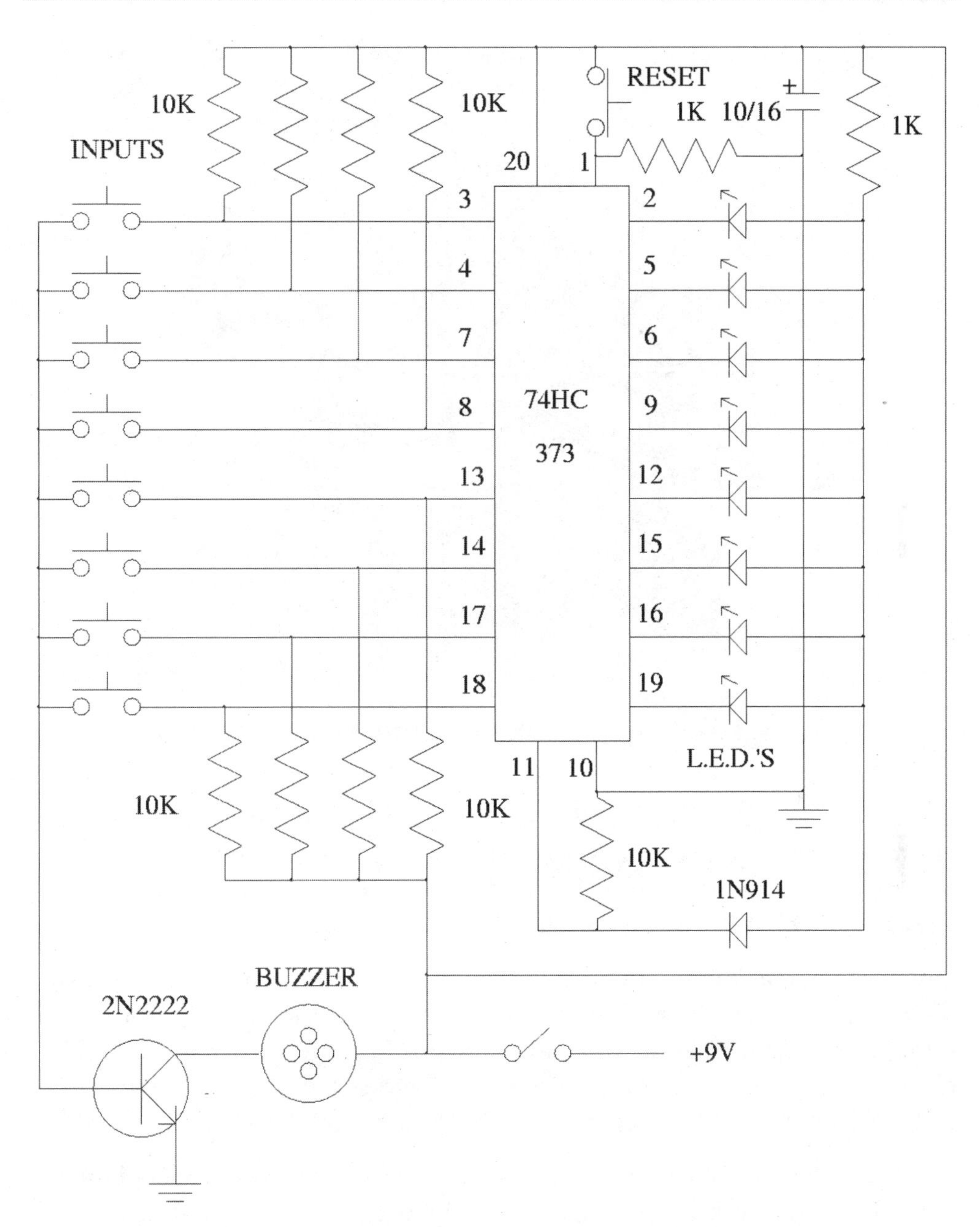

Figure 1-11: Battery-powered quiz machine

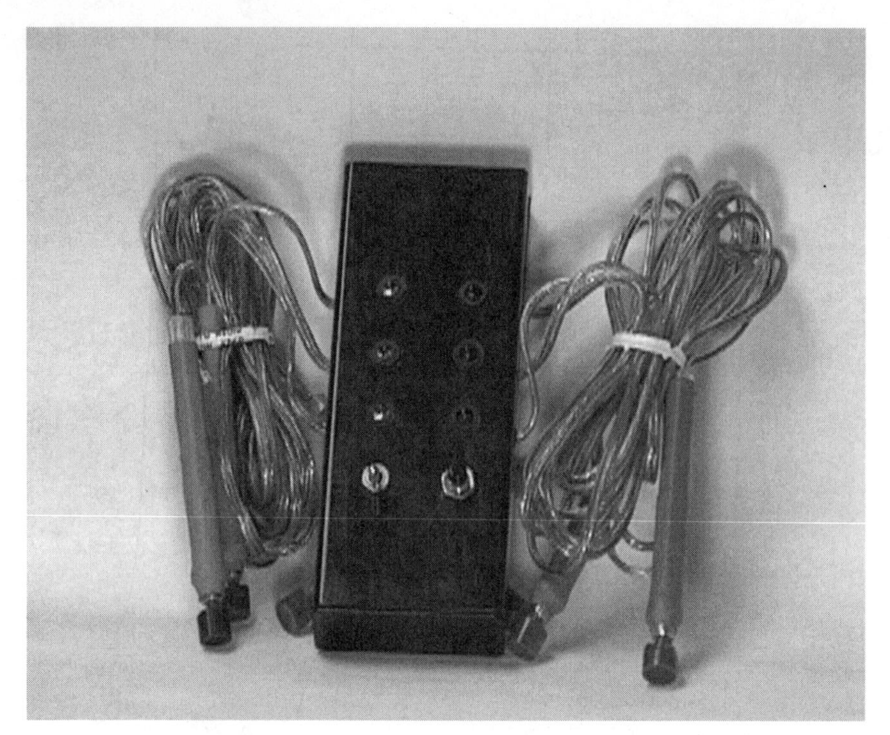

Figure 1-12: Picture of a battery-powered quiz machine

LEDs drop 1.6 volts. CMOS, at 9 volts, latches at around 2 volts, allowing proper latching without the two-transistor amplifier. TTL doesn't work because it requires a "zero" to be under 1.2 volts. The box used is about 3 inches by 2 inches by 1 inch tall.

I use a layout that has ten holes on the top of the box for eight players. For six players, use just eight holes. They are arranged in two rows. The wires from the switches come into holes on the sides of the box near their matching lights. The extra two holes on the top of the box are for the on/off switch and the reset switch. Make sure to leave room in the box for the 9-volt battery.

The 10K pull-up resistors can be used from a 10K resistor array soldered on top of the 74HC373. This arrangement saves a lot of space. The 1N914 diode in the latch circuit helps pull the latch 0.6 volts lower, to ensure proper latching. Otherwise, the latch sometimes fails when the battery is weak. This diode and the 10K resistor were only added in later designs.

The hand units for this model use BIC (or similar) pens that have stopped writing. They are cut in half and make 2 devices each. The switches are Radio Shack 275-1574 or equivalent. The wires are soldered to the switch and then the pen body is slid over the switch terminals. Glue and heat-shrink tubing firmly attach the switch to the pen body.

Quiz Machines: Plug-in Hand Units 1-12

The next version of the quiz machine offers high-power outputs and separate removable hand units. This makes possible LEDs in the control console as well as in the hand units. With this design you can even use 6-volt or 12-volt lights for more brightness. It can also drive opti-couplers to run 110-volt lights or relays.

The power source can be 6 to 12 volts from an AC adapter, since the IC is CMOS. With a 6-volt power source, I've used #1847, 6-volt, 150-mA lights with

1	9-volt DC–AC adapter
8	1N914 diodes (or equivalent)
9	momentary switches
1	SPST power switch
1	20-pin IC socket
1	74HC373
8	1N914
8	2N2907 transistors
1	box, 12 × 8 × 3 inches (approx.)
8	boxes, 3 × 3 × 1 inches (approx.)
1	circuit board, about 2 × 2 inches
1	buzzer
24	lights, #1847 or equivalent (or high-intensity LEDs)
24	light sockets (or LED holders)
8	1/4-inch stereo jacks
8	1/4-inch stereo plugs
48 feet	two-conductor shielded stranded wire (6 feet per player)
Miscellaneous	resistors and filter caps

Table 1-14: Parts list

no problems. They fit a Radio Shack 272-325 socket. Don't use the screw-in style lights, as they typically work their way loose during transportation.

The hand units are 3-inch-square boxes with lights on the front and back. They also have larger, heavy-duty "slappable" switches on top. A Radio Shack 275-609 or equivalent switch will work. The hand units connect to the base unit via

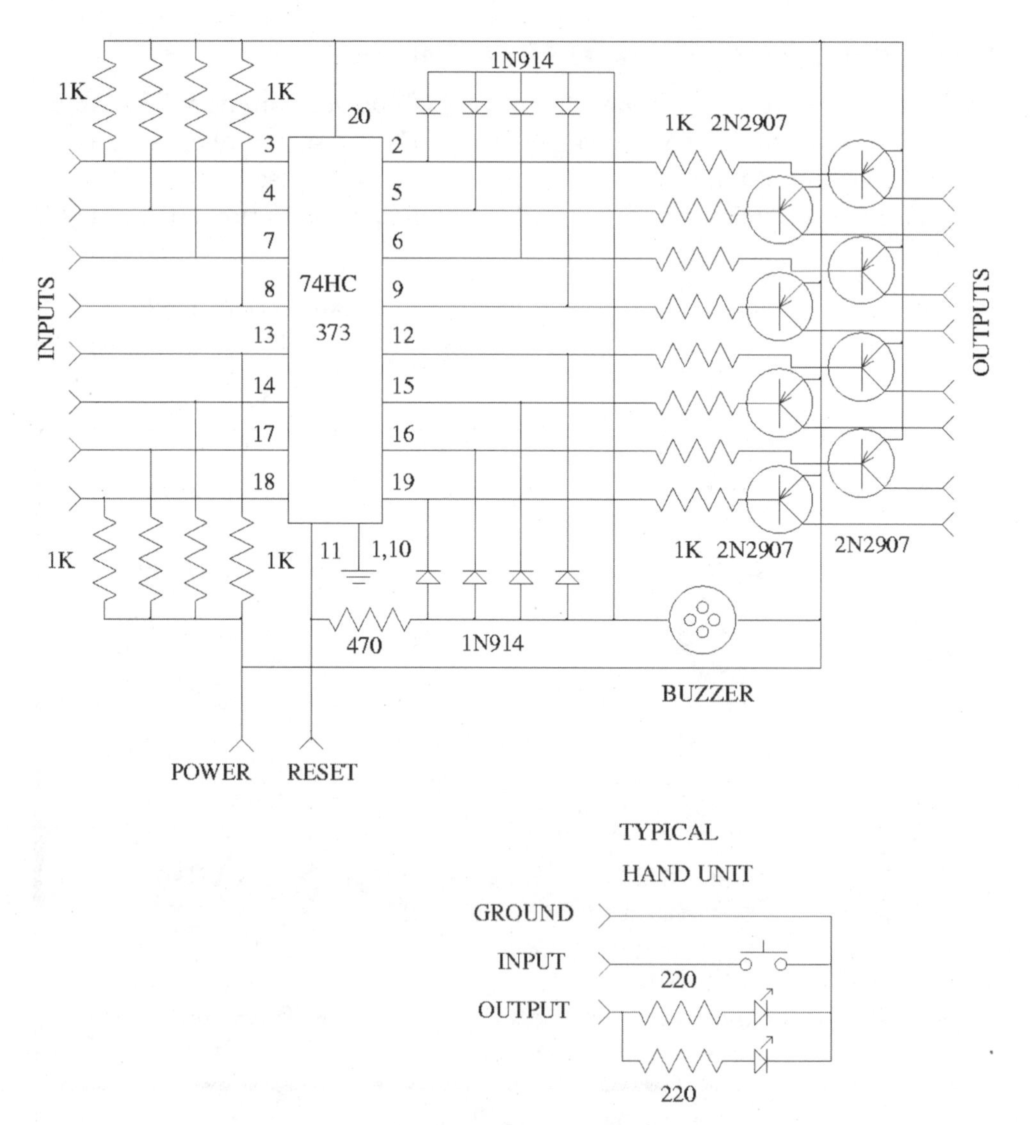

Figure 1-13: Quiz machine with plug-in hand units

1/4-inch "stereo headphone" type plugs. They can also connect in groups of four using 9-pin DIN (computer serial port) plugs and jacks.

The control console is 12 inches by 8 inches by 3 inches deep. A console with a slanted top is preferred. It can have power, reset, a buzzer switch, and eight lights on it. The lights are also on the hand units. A smaller control console will work, as there are only a few parts that go inside.

I've also made a version of this machine in a briefcase. The control console is in the top cover of the briefcase, so that when it opens up, the lights face out. The hand units can be set in the bottom of the briefcase.

A 555 timer could control a second buzzer and have it repeat every 10, 20, or 30 seconds, and this timing can be adjustable. This would tell if the question wasn't answered within an allotted time. The buzzer can be switched among no buzzer, buzz immediate, buzz on input and delay, and buzz on delay only.

Quiz Machines: Quiz Master 1-13

The last version is the "quiz master." It allows monitoring of up to 16 teams of 4 players each. That is 64 contestants! The schematic is configured for 12 teams of 3 players each, or 36 contestants, because that is the most requested arrangement.

The "quiz master" has two latches and a matrix of input switches that connect via several nine-pin plugs. Binary encoders encode the data and feed "or" gates that toggle the latch. The fastest team is represented with a letter of the alphabet programmed into a 74S188 PROM. The team's player is shown with a number, via a 74LS48 BCD to seven-segment decoder. So a winner might be displayed as "A 1," which means team A, player 1. This takes a little getting used to. Practice with the players before the game begins.

The bottom set of inputs, on the schematic, is actually divided up as two groups of three. The player's switches can plug in using two 9-pin connectors. You may want to use a larger 9-pin connector, as the wires from 18 players can get confusing. You could also use six 8-pin connectors and use only 7 pins of each to connect six players each. This would make soldering easier. Color-code the heat-shrink on the hand units for the teams, and then number them one to three for the player numbers.

The quiz master works on 9 volts DC from an AC adapter. Then, use a 7805 voltage regulator to provide the needed 5 volts. Don't forget to use a 100–470 µF filter capacitor on the power coming in and on the power leaving the voltage regulator. Then add 0.01 to 0.1 caps across the ICs to keep them from being bothered by any switching noise.

The parts layout is very similar to the interconnecting version and uses exactly the same box. On the front of the box, glue a red piece of plastic with beveled edges for the LED displays. On the back, mount the jacks for the hand units. The circuit board inside the box has a 22-pin edge connector on one end. The other end is cut off short enough to fit the board into the box. Then a short piece of circuit board is soldered on, perpendicular to the cut-off end, to hold

1	9 Volt DC–AC adapter, with LM7805 regulator
1	SPST power switch
2	20-pin IC sockets
2	16-pin IC sockets
3	14-pin IC sockets
1	8-pin IC socket
2	74LS373
2	74LS10
1	74LS27
1	74LS48
1	74S188
1	555 timer
2	common cathode LEDs
6	2N2222 transistors
1	box, 2 × 6 × 4 inches (approx.)
1	22-pin edge connecting circuit board, about 3 × 3 inches
1	buzzer
2	9-pin jacks
2	9-pin plugs
36	momentary switches
96	two-conductor stranded wire (6 feet per player)
12 feet	1/2-inch PVC pipe
Miscellaneous	resistors and filter caps

Table 1-15: Parts list

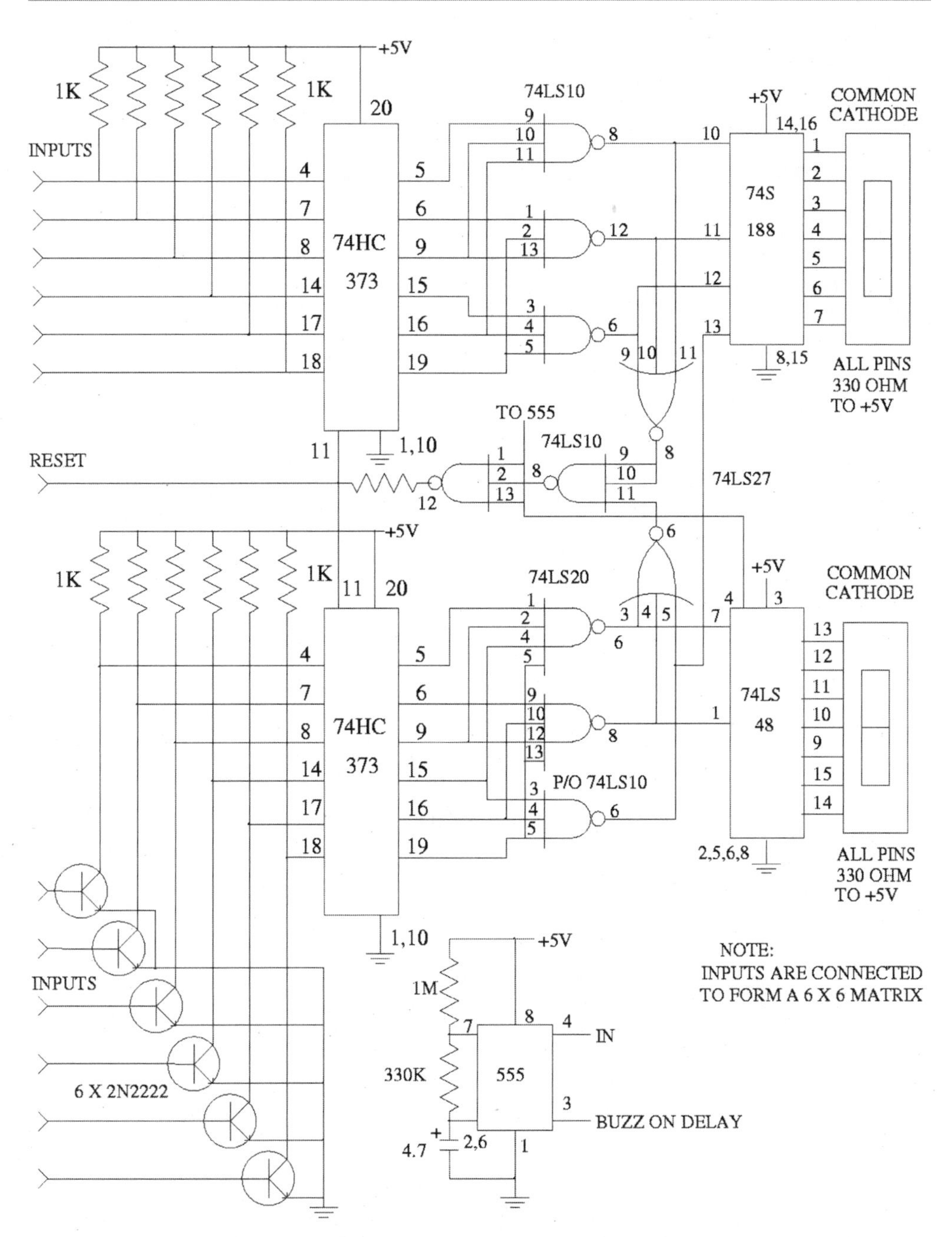

Figure 1-14: Quiz master

the LED displays. The 22-pin circuit board socket is mounted on the back above the jacks, so that the LED displays line up on the front.

Quiz Master EPROM Programmer 1-14

The quiz master needs a method of displaying letters on a seven-segment LED display. To my knowledge, no one makes a chip designed to do this. From my experience in making early computer BIOSs, I decided to use a fusible link PROM. One that's small enough isn't readily available, so a bigger one is used with the unused address pin tied high.

A 74S188 fusible link PROM can easily be programmed to do the job. All you need to do is set up the address and data (one bit at a time) according to Table 1-16 and press the program button ten times to program the PROM. Once it is programmed, plug it into the quiz master and test its ability to display letters. If a "1" isn't programmed, the segment won't light; reprogram that bit and address. If a "1" is accidentally programmed where it shouldn't be,

Symbol to Display	Segment XABC DEFG	Address
NONE	0000 0001	0000
A	0111 0111	0001
B	0001 1111	0010
C	0100 1110	0011
D	0011 1101	0100
E	0100 1111	0101
F	0100 0111	0110
NONE	0000 0001	0111
NONE	0000 0001	1000
G	0101 1110	1001
H	0011 0111	1010
J	0011 1100	1011
L	0000 1110	1100
P	0110 0111	1101
U	0011 1110	1110
NONE	0000 0001	1111

Table 1-16: Programming table

either scrap the chip, or change the status of A4 (pin 14) on the programmer and in the quiz master. Then try programming it again.

The PROM needs 10.5 volts to be programmed. I used a 12-volt AC adapter feeding a 7812 voltage regulator. Then two 1N4002 diodes drop that to 10.6 volts. A 7805 regulator provides the needed 5-volt source. Don't forget the filter caps!

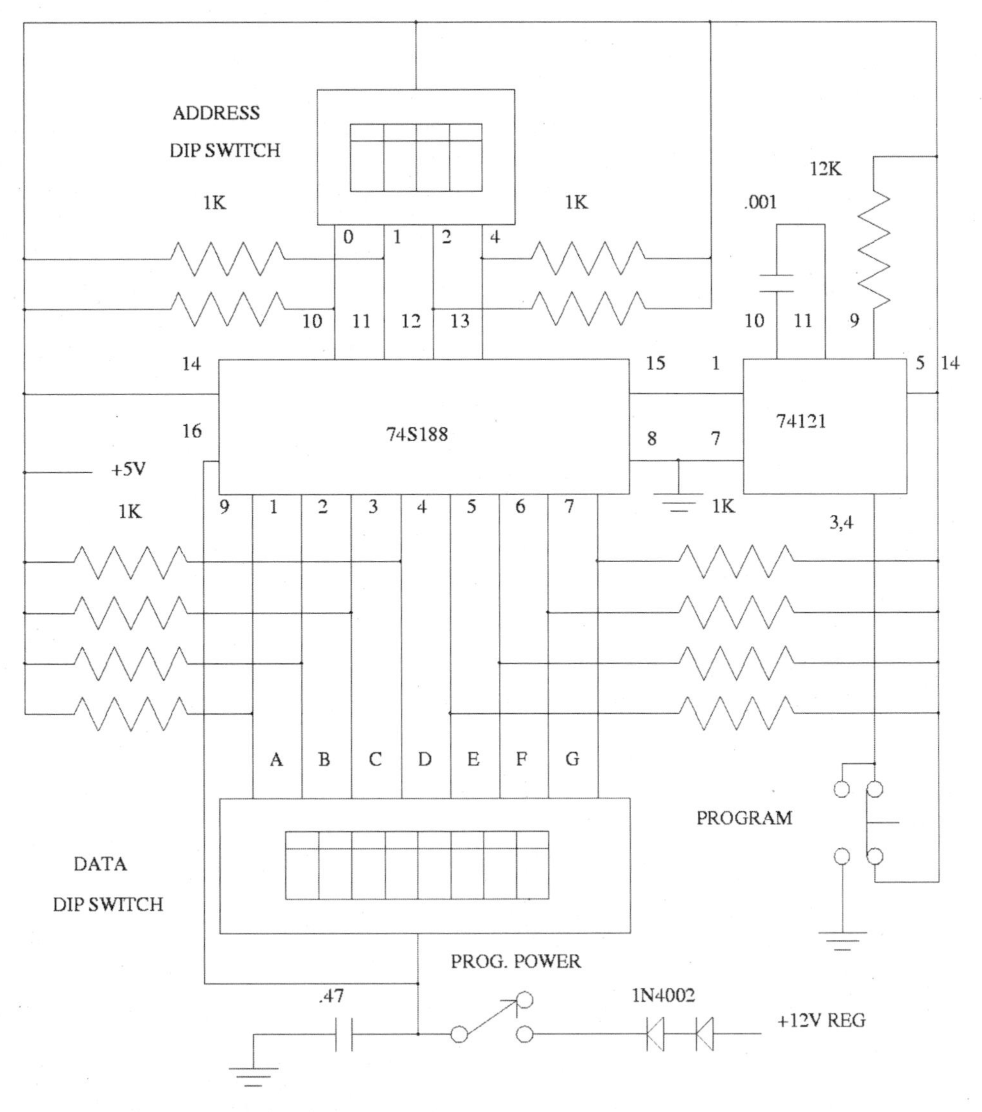

Figure 1-15: PROM Programmer for quiz master

1	12 Volt DC–AC adapter, with LM7812 regulator
1	SPST power switch
1	DPDT momentary switch
1	16-pin IC socket
1	14-pin IC socket
1	8-position DIP switch
1	4-position DIP switch
1	74LS121
2	1N4002 diodes
1	breadboard or circuit board, approx. 3 × 3 inches
Miscellaneous	resistors and filter caps

Table 1-17: Parts list

Letters such as "I" and "O" are skipped to avoid confusion with numbers. Other letters, such as "K," "M," and "N," are skipped because they're hard to make on a seven-segment display.

Follow the procedure carefully. Note that only one bit is programmed at a time. Take your time. These cannot be erased once they are programmed!

Procedure:

1. Select the address that is to be programmed.
2. Select a bit to be programmed to a "1" (only program one bit at a time).
3. Apply 10.5 volts via a jumper or switch.
4. Press the "program" button ten times.
5. Remove the 10.5 volts applied in step 3.
6. Go to step 2 until all bits for that address are programmed.
7. Go to step 1 until all addresses have been programmed.

Seven-segment display layout:

```
AAAAAA
F    B
F    B
F    B
 GGGGGG
E     C
E     C
E     C
 DDDDDD
```

FFD21 to LED Adapter 1-15

Some older Numerical Control equipment may still be in good working order. Getting needed replacement parts for the displays is not always easy. If you find the parts, they are enormously expensive. Some companies will even try to sell you a whole new display system, because it's cheaper than replacing the display devices. A simpler solution is to convert them so that they can use modern seven-segment LED displays.

The display that is easiest to replace is the FFD21. It uses filaments for its segments. These filaments wear out just as lights do, but perhaps more quickly, from vibration. To convert these to LEDs, all you need to do is add resistors in series. This limits the current. The filaments operate on 6 VAC, and the LEDs operate on 1.6 VDC. To reduce the current, use seven 220-, 330-, or 470-ohm resistors, depending on how bright you want the display to be. The AC doesn't seem to bother the LEDs, but a 1N4002 diode can rectify that and end your worries.

The biggest trick is squeezing it all together. Because of limited space, you might have to extend the front of the cabinet to make room for the LED display adapters. You need an LA-6460 type display, eight resistors, a diode, and

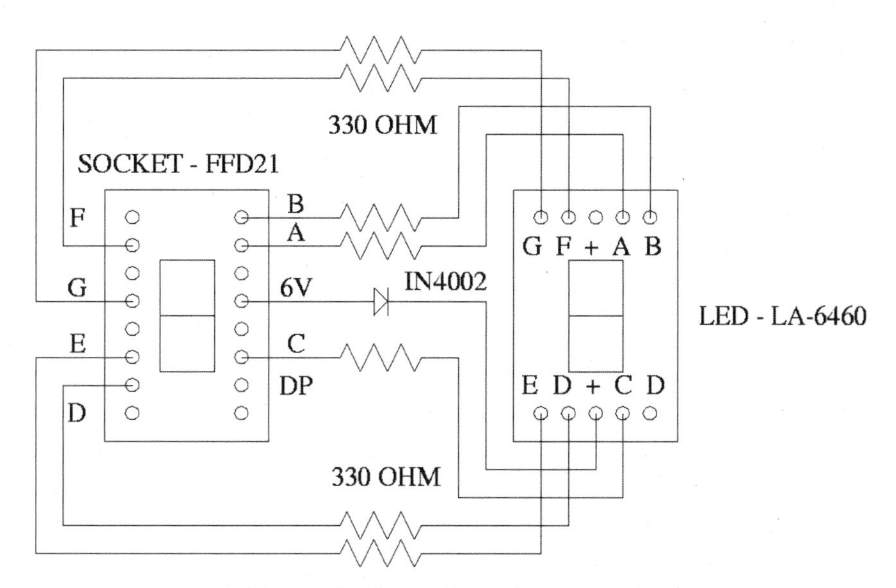

Figure 1-16: FFD21 to LED adapter

1	LA-6460 LED display
8	220-ohm resistors
1	1N4002 diode
1	16-pin header

Table 1-18: Parts list

a 16-pin header. Figure 1-16 doesn't show the resistors, but they are inserted on each wire as it goes from the header to the LED display.

All wiring views are from the top even though you can't see the pins from the top. The resistors are soldered onto the LED display, their leads are bent so they line up with the correct pins on the header, and then they are soldered in.

NIXIE Tube and 841 to LED Adapters 1-16

The next adapter is primarily for NIXIE tubes. It will also work with any display that works with a 7441 driver. NIXIE tubes have 170 volts on them. Their normal driver IC, a 7441, has an incompatible pinout. This driver IC must also have a socket. If it doesn't, you need to remove it and add a socket. Removal is easy—cut off the leads at the IC, then unsolder the leads one at a time. A header plugs into that socket and small wires go to the new LED circuit.

The LED display fits over a 74LS47 onto a small circuit board. The circuit board should be extra tall so it's easy to fasten it in place. Use 220-, 330-, or 470-ohm resistors to reduce the load on the power supply. The smaller resistors give brighter displays, but may overload the power supply. Stand each resistor on end for each segment. This uses less room on the adapter board. If you are replacing an entire row of NIXIE tubes, use a circuit board big enough to hold all the LED displays and drivers.

Some machines have 5-volt power supplies that may not be able to handle the added current for the LED displays. The NIXIE tubes used 170 volts, so the power supply was aimed more at voltage than current. You might even have to use 470-ohm resistors to reduce the current. If that doesn't help, you may have to add an external power supply. Other options are to use even larger resistors or beef up the internal power supply.

If you have to use larger resistors and the display is too dim, try adding a "visor" to the display. The visor only needs to stick out 2 to 4 inches to block the overhead lights and can be fastened on with two screws.

Sometimes the power supply seems to work at first, but overheats later. Instead of taking apart all the adapters and changing the resistors, it is possible

1	16-pin IC socket or header
1	74LS47
1	common anode LED display
7	220-, 330-, or 470-ohm resistors (see text)
1	circuit board, approx. 1 × 2.5 inches

Table 1-19: Parts list

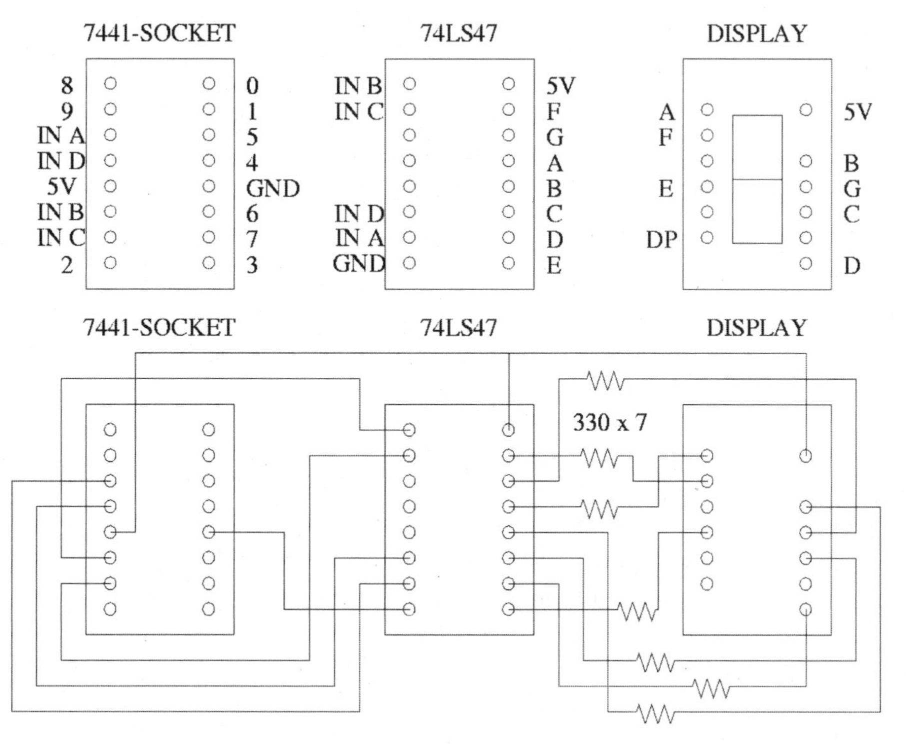

Figure 1-17: NIXIE tube to LED adapters

to add a 1N4002 diode or two in series with the power supply and running to the LED display. Each diode drops the power by another 0.6 volts. This change will also make the display dimmer, but will reduce the load on the power supply as well.

When retrofitting an entire display unit, use a circuit board big enough to hold all the LED displays, drivers, and resistors and fasten them in place. Then you need to run just four wires out to each socket or header.

Multiply quantities by the number of NIXIE tubes being replaced.

8052 Basic Energy Management 1-17

This is one circuit that I certainly did not invent. There have been many other projects on how to use the 8052AH-Basic. The problem is that with all of the articles I've read on the 8052, not one really puts it into practical use. Most do little more than flash lights! This application of the 8052 has been used for years to run a heating system and allow easy monitoring.

The 8052AH-Basic is an eight-bit processor with built in ROM, RAM, and an 8-bit I/O port. The ROM has a small version of BASIC programmed into it. The RAM is too small for practical purposes, so an external RAM chip is almost always used.

One of the big problems with a hot-water baseboard system is that it "over-shoots" the desired temperature. This happens regularly in the spring and fall, when it isn't as cold outside. When the thermostat senses that the room is up to the temperature set, the furnace turns off. But this leaves the heaters full of 180-degree water. That causes the temperature to rise higher than the thermostat, and the occupants get used to it. They might even open the windows to cool off. They then feel cold when it gets down to normal and wonder if the furnace quit.

The problem of overshoots can be solved without a programmed controller. There are discrete analog devices out there to do the trick, but it takes two of them at over $100 each to do the job. The 8052-based controller can also add other features as time goes by—features such as an alarm to let you know if the heat really does fail.

To make a furnace controller work, I need to know the outside temperature, the boiler temperature, and the room temperature. So what is needed is an 8052, with memory and a multichannel analog-to-digital converter. Outputs are needed to shut off the boiler, to display the temperatures, and to sound an alarm.

Sample Program

```
10     REM THREE TEMPERATURE DISPLAY & ALARM
15     REM BY BOB DAVIS IN 8052AH-BASIC
20    XBY(0E000H)=0:  REM SELECT SENSOR #1
25    C=.98*(XBY(0E000H))-273+125 :  REM GET TEMP IN DEGREES C
30    T1=9/5*C+32 :  REM CONVERT C TO DEGREES F
35    XBY(0E001H)=0
40    C=.98*(XBY(0E000H))-273+125
45    T2=9/5*C+32
50    XBY(0E002H)=0
55    C=.98*(XBY(0E000H))-273+125
60    T3=9/5*C+32
65    S=PORT1-INT(PORT1/4)*4
70     IF S=1 THEN F=T3 :  REM DOWN IS OUTSIDE
75     IF S=3 THEN F=T1 :  REM 3 IS MIDDLE - ROOM
80     IF S=2 THEN F=T2 :  REM UP IS BOILER
85    R=ABS(F-INT(F/10)*10) :  REM COMPUTE RIGHT DIGIT
90    L=ABS(INT(F/10)-INT(F/100)*10) :  REM COMPUTE LEFT DIGIT
95    XBY(0C000H)=R+L*16 :  REM DISPLAY TEMPERATURE
100    IF F>99 THEN M=0 ELSE M=4 :  REM MSB DISPLAY
105    IF F<0 THEN N=0 ELSE N=8 :  REM NEGATIVE TEMPS
110    IF F<50 THEN A=16 ELSE A=0 :  REM SOUND ALARM
115    IF T2+T3>210 THEN O=32 :  REM BOILER+OUT OVER 210
120    IF T2+T3<190 THEN O=0
125   PORT1=3+M+N+A+O :  REM 3 IS TO READ SWITCH
200    GOTO 20
```

Programming is a little tricky at first. When you power up the 8052AH-Basic, it first tries to determine the communications rate. The first thing it must receive is a space bar. That sounds a little strange, but you plug it in and from Procomm or Windows terminal, you press the space bar. Then you can type in the program just given, or have Windows terminal do it for you. (See the next section on the 68HC11.)

The analog input is selected by the address that is written and found in program lines 20, 35, and 50. The input is then adjusted so that a full-scale reading of 255 is equal to 2.5 volts, a factor of 0.98. Then it is converted to degrees Celsius by subtracting kelvins (–273) and adding 125, for the 1.25-volt bias on the negative reference. This is then converted to degrees Fahrenheit by the usual formula. The status of the display select switch is then read, and the desired temperature is displayed.

The temperature has to be converted from binary to decimal to be displayed. The left digit is multiplied by 16 to show up in the higher 4 bits for the left display. That information is then sent to the 4511s. Then the information is put together for port 1. A 4 is used to turn on the "1" for temperatures over 99 degrees. An 8 is used to turn on the negative sign. Line 110 adds a 16 to sound the alarm if the displayed temperature drops below 50 degrees.

Qty	Item
1	40-pin socket
2	28-pin sockets
1	20-pin socket
1	16-pin socket
1	14-pin socket
1	8052AH-Basic available from Jameco
1	6264 8K by 8-bit memory chip
1	ADC0808 eight-channel analog-to-digital converter
1	74LS02
1	74LS138
1	74LS373
1	4.9-MHz crystal
3	NSN64R common cathode LED displays
4	2N3906 transistors
1	LM336 2.5-volt reference
3	LM335 precision temperature sensors
1	three-position switch
1	circuit board, approx. 3 × 4 inches
1	9-volt DC–AC adapter, with a 7805 voltage regulator
Miscellaneous	resistors and capacitors
Optional setup for remote RS-232 monitoring:	
1	Max232
1	16-pin socket
4	10-µF/16-volt caps
Optional setup for remote RS-422 monitoring:	
2	75176
2	4-pin sockets

Table 1-20: Parts list

If the combined temperatures of the boiler and the outside temperatures exceed 210 degrees, the boiler is throttled back with a "32." A 20-degree differential keeps the boiler from short cycling. Port 1 then outputs the results of the four decisions and a "3" so that the switch position can be read.

The display has a tendency to jump 2 degrees at a time because of the conversion formulas that are used. That is accurate enough for an alarm and to control the boiler. Reducing the two-degree jump is not easy. The input range of 2.5 volts is close to the 0 to 255 range that the computer sees, and changing it

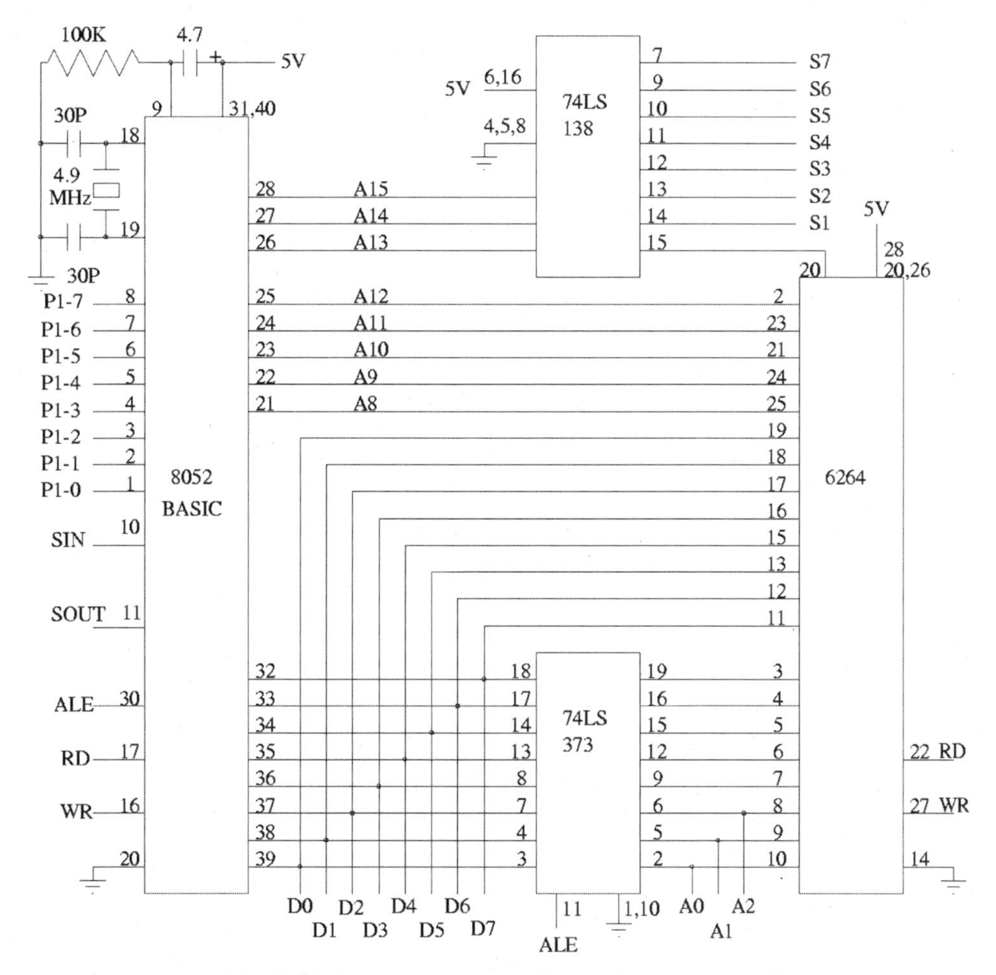

Figure 1-18: 8052 Energy Management Controller CPU section

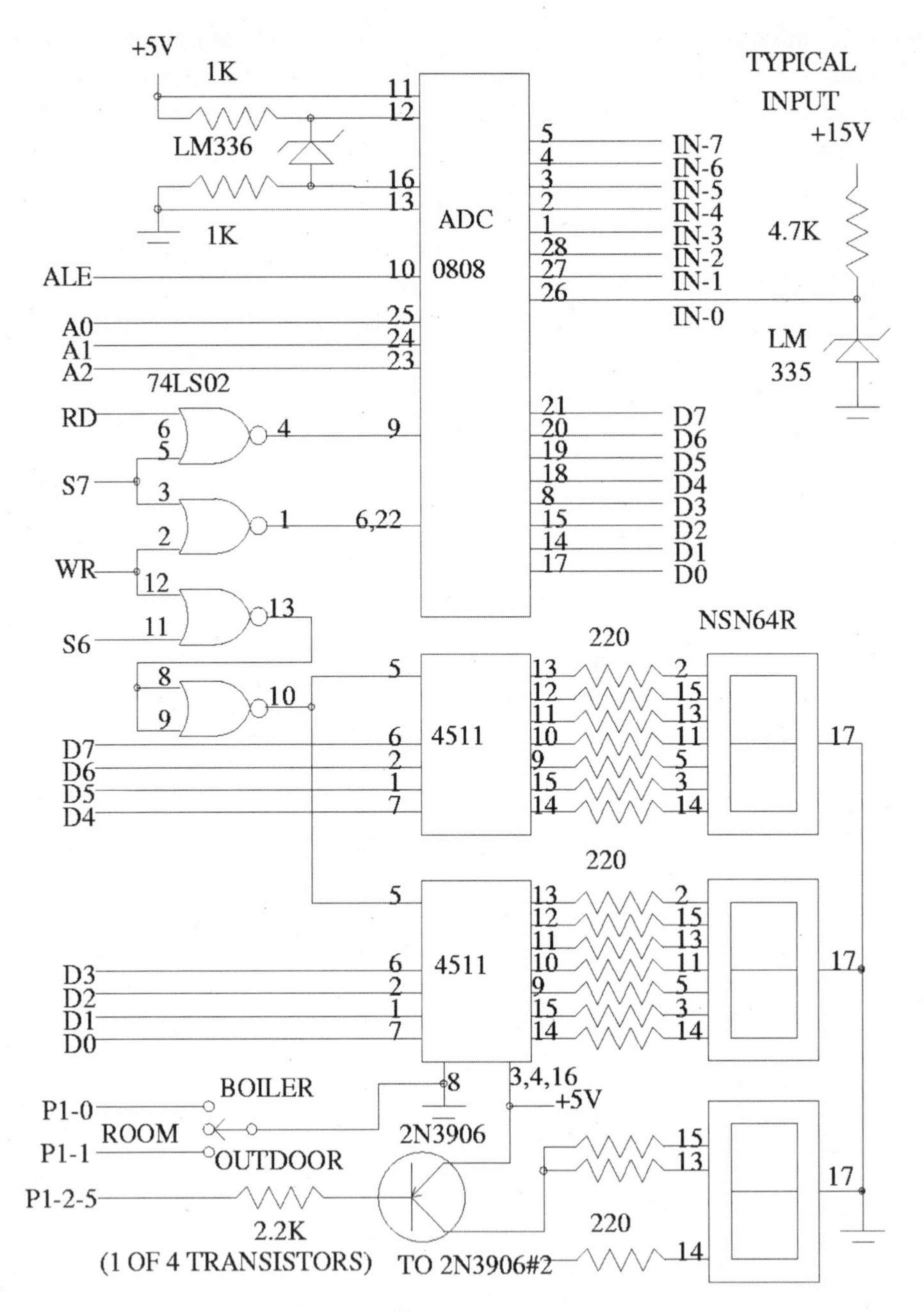

Figure 1-19: 8052 Energy Management Controller analog section with LED display

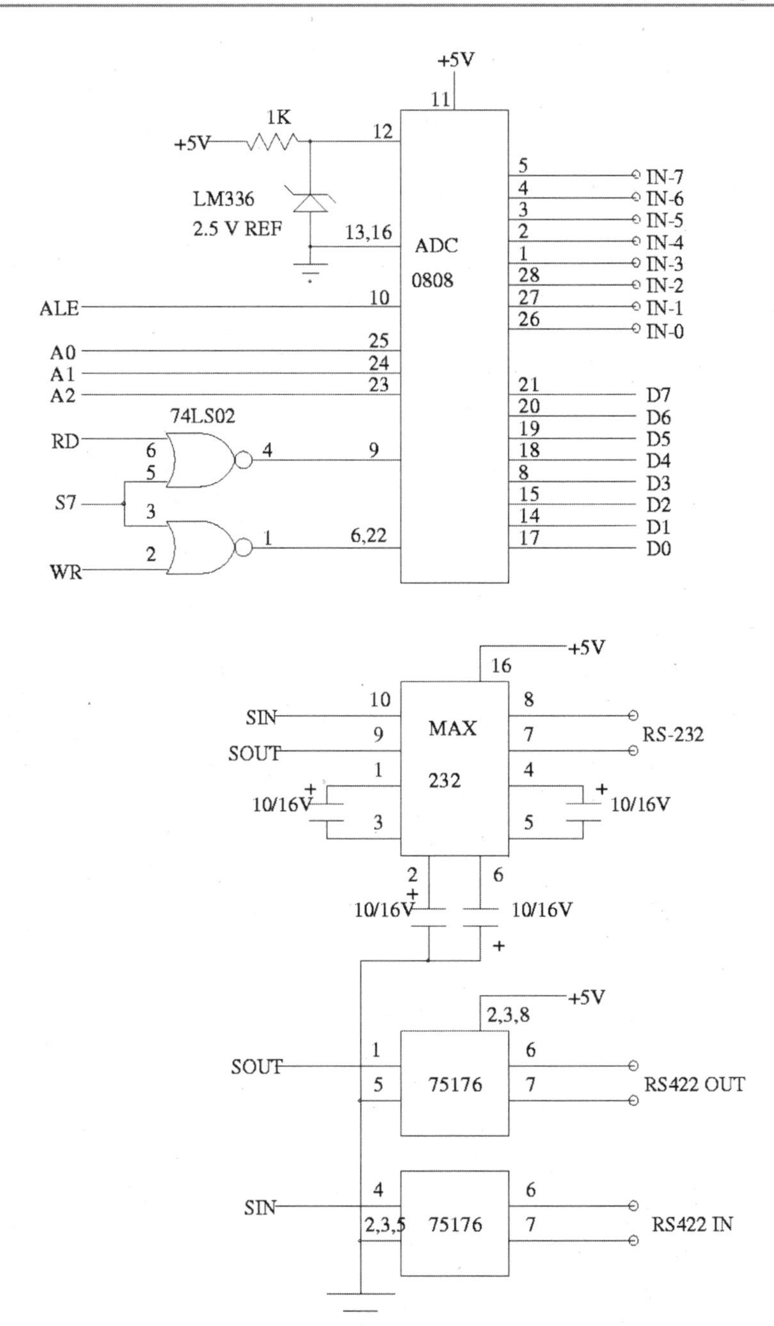

Figure 1-20: 8052 Energy Management optional analog section with serial I/O

might be tricky. The sensors could be changed to improve sensitivity instead. With the monitor set up as shown, the temperature range is about –199 to +199 degrees F. Another solution would be to use LM34 temperature sensors. That would limit the range to 0 to 255 degrees. The resulting temperature would step only 1 degree at a time.

There are two schematics for the analog section. Figure 1-19 includes an LED display for easy local monitoring. The software listing is for that version. The second schematic, Figure 1-20, has both an RS-232 and an RS-422 interface for easy remote monitoring. This version would work better for a remote IBM PC-controlled system.

At the bottom of the first schematic for the analog section, there is only one 2N3906 transistor. In actuality, there are four transistors. The one shown is used

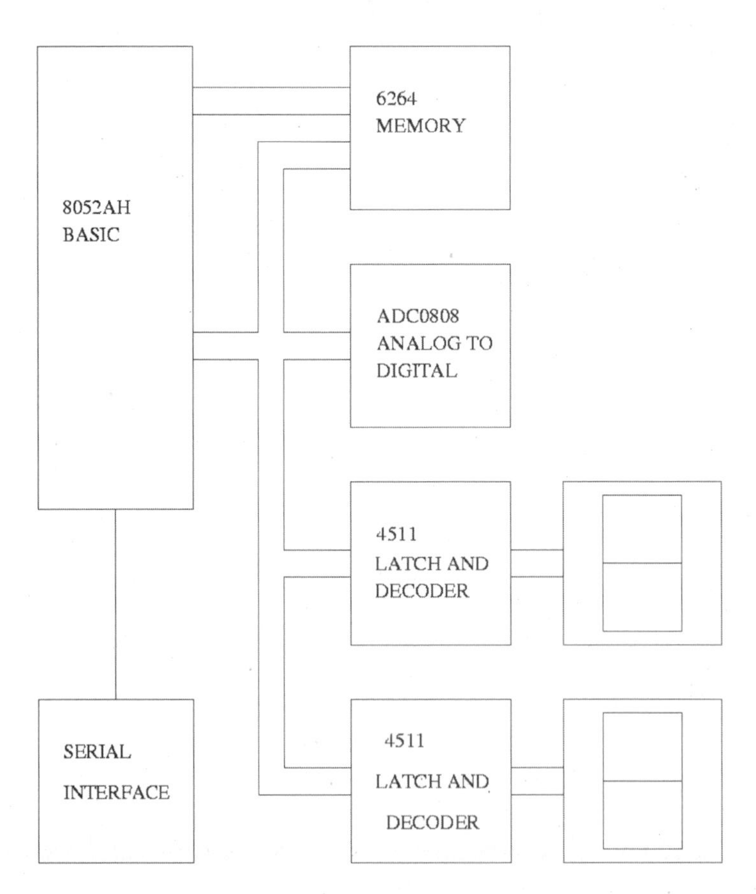

Figure 1-21: 8052 Energy Management Block diagram

to light up a "1" for temperatures over 99 degrees. The next one lights up a "–" for temperatures below 0 degrees. The third transistor sounds a buzzer if the selected temperature drops below 50 degrees. The last one can run a relay to shut off the furnace if it is deemed to be too hot. The base of each transistor is fed from the 8082s port 1, bits 2, 3, 4, and 5. Bits 0 and 1 are used for the "select temperature to display" switch.

68HC11 Energy Management System 1-18

The 68HC11 offers many advantages over the older 8052. The most important one is that it has a built-in analog-to-digital converter. It also has built-in RAM and programmable EEPROM. It even features several built-in parallel input and output ports. Besides all that, I was given a pile of them to play with.

When researching an energy management system, we considered several options. You can put an ordinary IBM PC in each building's furnace room, or a specialized PLC (programmable logic controller) could be used. The best option was a standard microprocessor. The PC is cheap, and we have piles of donated 386s. They would not last long, however, in a furnace room with the high range of temperatures, dust, and general abuse. Besides that, where do you put the monitor? Nor do they come with analog-to-digital converters, so one would have to be added. A dedicated PLC is expensive, proprietary, and not everyone has the software or the knowledge to program them. Most PLCs once again do not come with the needed analog-to-digital converter.

A standard microprocessor is ideal. The 68HC11 requires virtually no support chips, except a relay driver, and RS-232 or RS-422 drivers. Then, I was surprised that no one seemed to make a suitable printed circuit board. I wanted eight analog inputs, eight digital inputs, and eight relay outputs. So once again, I designed my own circuit board.

The model used here is a 68HC11F. It offers more parallel ports than the more common "E" version. Besides that, I was also given a pile of these. The software was written so that it should work as well on the "E" model. However, the schematic and circuit board are designed for the "F" version and would need some real modification for the "E" version.

Then there is the problem of programming the 68HC11. The "F" version of the chip seemed to be a problem—finding a programmer for it is not easy. We settled on one from New Micros Inc. They are at 1601 Chalk Hill Rd., Dallas TX. 75212. Their phone number is 214-339-2204. The model chosen was the NMIS-0024. It has TTL serial ports so a small circuit board with a MAX232 had to be added to communicate with it. I also added DIP switches and LEDs to allow simulation of the completed project.

To program the 68HC11, I used Windows 3.1's "Terminal" emulator. Select "settings" and "communications," then select "9600" to set the required communications rate. Plug in the programmer and it starts with the "Buffalo" monitor. You type "ASM 0E00" and type in the lines in the listings. Or better yet, edit the file, removing the line numbers and the comments. Then type "ASM 0E00" and select "Settings," "text transfer," "character at a time," and "delay between characters." Then Select "Transfers" and "send text file." Enter the name of the edited file, and Terminal will program the chip for you.

After programming the 68HC11, a start vector must be programmed into the highest 2 bytes of the EEPROM. This is done in Windows terminal. Enter "MM 0FFE" to modify address 0FFE. Then when it says "FF" enter "FE" and return. It will then say "FF" again, and you enter "00." Then, exit memory modify mode and you have successfully programmed a jump to "FE00" into the highest bytes of the EEPROM. When powered up in the Universal controller, the 68HC11 will jump to FE00, the beginning of the EEPROM, and start executing from there.

It is possible to use just one telephone line to talk to the 68HC11, but I chose to use two. We own our phone lines around the campus where I work, so that is not a problem. Using two lines also makes it easy to check on the 68HC11 with Windows Terminal or Procomm. Otherwise a BASIC or C program would be needed to select send or receive mode for the PC's serial port.

To access the 68HC11, just press Return, the address, Return, a value for the relays, and Return. The 68HC11 then echoes the values in the eight analog-to-digital converters in degrees Fahrenheit and the contents of the 8-bit digital input port. It will also set the relays according to the value you gave it.

The software allows the use of ASCII codes—for example, typing address "2" on the keyboard will select 68HC11 device number 2. This is done by "OR"ing 30 when sending to convert the output to ASCII. When receiving it only looks at the lower 4 bits.

1	69-pin socket
1	16-pin socket
1	14-pin socket
2	8-pin sockets
1	68HC11F microprocessor
1	74LS05
1	ULN2003
2	75176
1	8-MHz crystal
1	four-position DIP switch
4	LEDs
1	LM7805 5-volt regulator
1	LM336 2.5-volt reference
8	LM335 precision temperature sensors
8	12-volt relays, miniature DIP style
1	circuit board, approx. 4 × 5 inches, or printer circuit board
1	12-volt DC–AC adapter
50	terminal strip positions (these come in combinations of 2, 3, 4, 5, or 6 at a time. See Figure 1-25 for more help.
Miscellaneous	resistors and capacitors

Table 1-21: Parts list

The software also converts the results to degrees Fahrenheit. This is made easier by first setting the analog-to-digital converter's reference to 2.55 volts. This is done with a 24K to 100K resistor to trim the reference voltage. The value needed may vary with the regulator. Test your regulator and change the resistor if it is needed.

```
REM ** 68HC11 EMS DOWNLOAD FILE **
ASM 0E00
0E00 LDX   #$1000      >  I/O ADDRESS
0E03 LDAA #$30         >  9600 BAUD
0E05 STAA $2B,X        >
0E07 LDAA #$00         >  8 BITS
0E09 STAA $2C,X        >
0E0B LDAA #$0C         >  TRANSMIT & RECEIVE
0E0D STAA $2D,X        >
0E0F LDAA #$93         >  TURN ON A TO D CONVERTER
0E11 STAA $39,X        >
```

```
0E13 LDAA #$F0         >  SET 1/2 PORT A AS OUTPUT
0E15 STAA $01,X        >
0E17 LDAA #$00         >  CLEAR STATUS LED'S
0E19 STAA $00,X        >
0E1B BRCL $2E,X $20 $0E1B  >  WAIT FOR INPUT BYTE
0E1F LDAA #$10         >  SET I/O ACTIVE LED
0E21 STAA $00,X        >
0E23 LDAA $2F,X        >
0E25 CMPA #$0D         >  CHECK FOR CARRIAGE RETURN
0E27 BNE  $0E17       >   RESTART IF NOT CR
0E29 BRCL $2E,X $20 $0E29  >  WAIT FOR INPUT BYTE
0E2D LDAA $00,X        >  GET ADDRESS SELECT SWITCHES
0E2F ANDA #$07         >  MASK OUT BITS 0-2
0E31 ORAA #$30         >  MAKE INTO ASCII
0E33 CMPA $2F,X        >  COMPARE TO INPUT BYTE
0E35 BNE  $0E17       >   RESTART IF NOT SAME
0E37 BRCL $2E,X $20 $0E37  >  WAIT FOR INPUT BYTE
0E3B LDAA $2F,X        >
0E3D CMPA #$0D         >  CHECK FOR CARRIAGE RETURN
0E3F BNE  $0E17       >   RESTART IF NOT CR
0E41 BRCL $2E,X $20 $0E41  >  WAIT FOR INPUT BYTE
0E45 LDAA #$30         >  SET SELECTED STATUS LED
0E47 STAA $00,X        >
0E49 LDAA $2F,X        >  SET PORT B RELAYS
0E4B STAA $04,X        >
0E4D LDAA #$10         >  SELECT A TO D PORTS 0-3
0E4F STAA $30,X        >
0E51 LDAA $30,X        >
0E53 BPL  $0E51       >   WAIT FOR A TO D CONVERTER
0E55 LDAA $31,X        >  STORE RESULTS OF CONVERSIONS
0E57 STAA $0200        >
0E5A LDAA $32,X        >
0E5C STAA $0201        >
0E5F LDAA $33,X        >
0E61 STAA $0202        >
0E64 LDAA $34,X        >
0E66 STAA $0203        >
0E69 LDAA #$14         >  SELECT A TO D PORTS 4-7
0E6B STAA $30,X        >
0E6D LDAA $30,X        >
0E6F BPL  $0E6D       >   WAIT FOR A TO D CONVERTER
0E71 LDAA $31,X        >  STORE RESULTS OF CONVERSIONS
0E73 STAA $0204        >
0E76 LDAA $32,X        >
0E78 STAA $0205        >
0E7B LDAA $33,X        >
0E7D STAA $0206        >
0E80 LDAA $34,X        >
0E82 STAA $0207        >
0E85 LDAA $06,X        >  GET INPUT PORT C
```

```
0E87 STAA $0208      >
0E8A LDAA #$70       >  SET SENDING DATA STATUS LED
0E8C STAA $00,X      >
0E8E LDY  #$01FF     >  SET OUTPUT COUNTER
0E92 INY             >
0E94 CPY  #$0209     >  SEE IF DONE OUTPUTTING
0E98 BEQ  $0EDD      >  BRANCH TO SEND CR AND LF WHEN DONE
0E9A LDAB $00,Y      >  GET DATA TO SEND
0E9D LDAA #$00       >
0E9F LDX  #$000A     >  DIVIDE BY 10 - A IN HEX
0EA2 IDIV            >
0EA3 XGDX            >
0EA4 LDX  #$000A     >  DIVIDE BY 10 AGAIN
0EA7 IDIV            >
0EA8 XGDX            >
0EA9 ORAB #$30       >  CONVERT TO ASCII
0EAB TST  $102E      >  WAIT FOR READY TO SEND
0EAE BPL  $0EAB      >
0EB0 STAB $102F      >  SEND RESULTS - BYTE 1
0EB3 XGDX            >
0EB4 ORAB #$30       >  CONVERT TO ASCII
0EB6 TST  $102E      >  WAIT FOR READY TO SEND
0EB9 BPL  $0EB6      >
0EBB STAB $102F      >  SEND RESULTS - BYTE 2
0EBE LDAB $00,Y      >  GET DATA TO SEND - AGAIN
0EC1 LDAA #$00       >
0EC3 LDX  #$000A     >  DIVIDE BY 10
0EC6 IDIV            >
0EC7 ORAB #$30       >  CONVERT TO ASCII
0EC9 TST  $102E      >  WAIT FOR READY TO SEND
0ECC BPL  $0EC9      >
0ECE STAB $102F      >  SEND RESULTS - BYTE 3
0ED1 LDAA #$20       >  SEND SPACE BETWEEN VALUES
0ED3 TST  $102E      >  WAIT FOR READY TO SEND
0ED6 BPL  $0ED3      >
0ED8 STAA $102F      >  SEND SPACE - BYTE 4
0EDB BRA  $0E92      >  GET NEXT DATA TO SEND
0EDD LDAA #$0D       >  DONE - SEND CR & LF
0EDF TST  $102E      >  WAIT FOR READY TO SEND
0EE2 BPL  $0EDF      >
0EE4 STAA $102F      >  SEND CR
0EE7 LDAA #$0A       >  LINE FEED
0EE9 TST  $102E      >  WAIT FOR READY TO SEND
0EEC BPL  $0EE9      >
0EEE STAA $102F      >  SEND LF
0EF1 TST  $102E      >  MAKE SURE LAST BYTE WAS SENT
0EF4 BPL  $0EF1      >
0EF6 JMP  $FE00      >  RESTART FROM BEGINNING
0EF9 STX  $FFFF      >  7 SPARE BYTES OF 256 BYTES ROM
0EFC STX  $FFFF      >
```

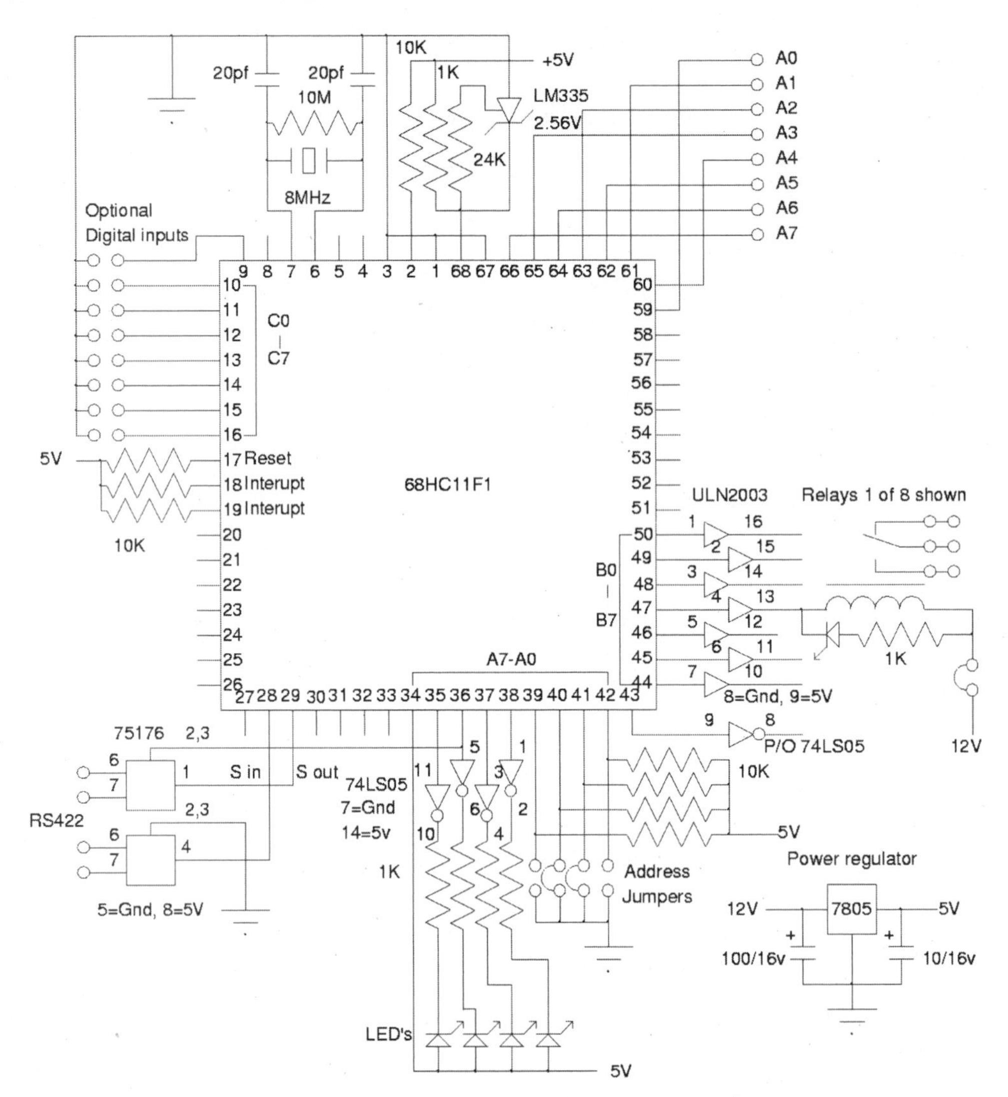

Figure 1-22: 68HC11 Universal Energy Management Controller schematic

An LM34 temperature sensor is used; it gives the temperature in Fahrenheit as volts. The voltage changes at 0.01 volts per degree. For example, 70 degrees is 0.70 volts. Since the 68HC11 sees an 8-bit value of 0 to 255, that represents 0 to 2.55 volts. Hence, 0.7 volts returns a value of 70. But that "70" is in binary—to convert to decimal you have to divide it by 10, or "A" hexadecimal. This division is done three times, once for each digit to be extracted. But the digits extracted will

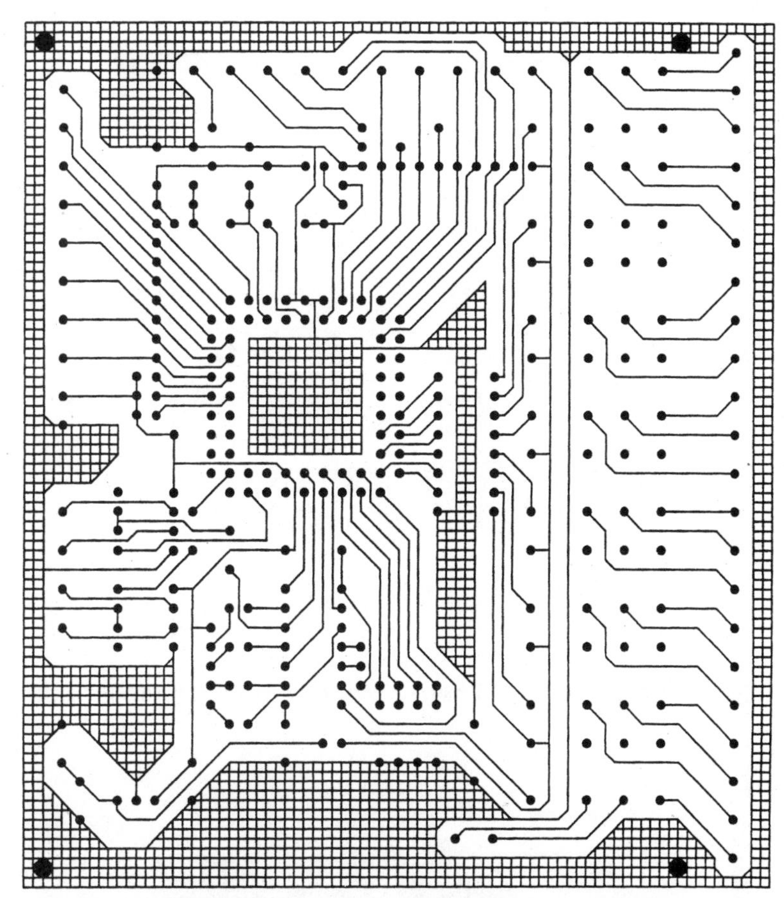

Figure 1-23: 68HC11 Universal Energy Management Controller circuit board

be backwards because you are looking at the remainder each time. That is to say, 072 degrees will appear as 270 degrees. At first I was very confused about this response. So instead, you must first divide twice to get the first and second digit. Then restore the original value and divide once for the third digit.

Four LEDs are also provided. They tell first if data is being received, then if the correct address has been received. The third tells that data is being sent back to the computer in response. The fourth LED wasn't used by the software.

A ULN2003 seven-line relay driver is used for the output port drivers. The ULN2003 is short a bit so I used one gate of the 74LS05 for the eighth bit. A ULN2803 eight-bit driver could have been used, but they aren't as common.

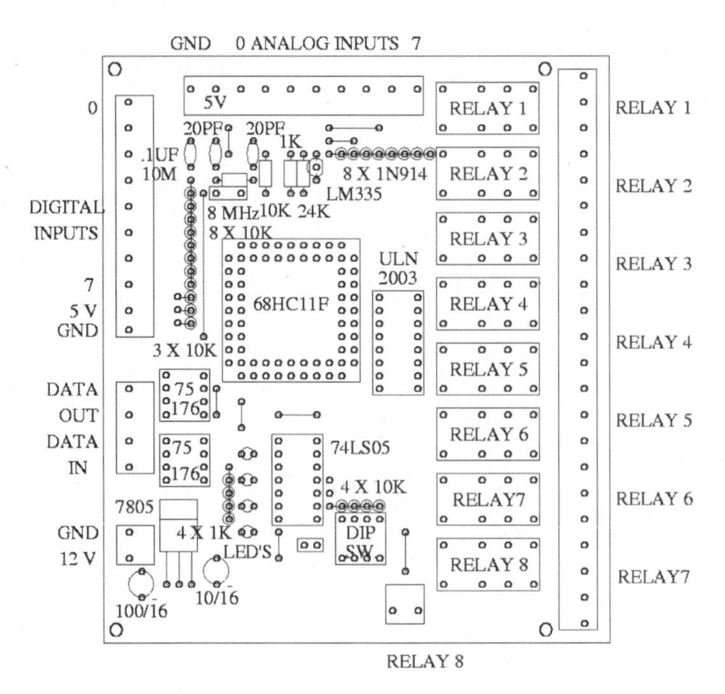

Figure 1-24: 68HC11 Universal Energy Management Controller parts layout

Figure 1-25: Picture of 68HC11 Universal Energy Management Controller

The relays have a jumper to turn them off. This is for an optional "override" switch. The switch could also signal a bit on the digital input to let the controlling computer know that someone has overridden the controller.

The printed circuit board has room for eight pull-up resistors on the digital input bits. A 10K in-line resistor array will work, or make one yourself with eight 10K resistors. There is also room for eight diodes on the analog input pins. These can be 1N914, or you can use 5-volt Zener diodes for added protection.

Two-Port RS-232 to RS-422 Converter 1-19

I've seen schematics of RS-232 to RS-422 and RS-485 converters, but they only handle one port. Some use the serial port RTS (Ready To Send) to toggle the direction of the RS-422 port. That wastes up to half of the MAX232. These RS-232 to RS-422 converters are not cheap, usually costing over $50. So I designed one to use both serial ports.

For my energy management system, I wanted to split the campus in half, with one serial port handling devices on the east side of campus, and one handling devices on the west side. At 16-68HC11 devices possible per RS-422 port, that means that I could access up to 32 devices. Hence, a two-port version was needed.

RS-422 offers several advantages over RS-232 for serial communications. RS-422 is "differential." That means that every signal uses two wires, so that any noise induced on one line is canceled on its pair. RS-422 also uses normal 5-volt logic

Qty	Part
1	16-pin socket
4	8-pin sockets
1	MAX232
2	75176
1	LM7805 5-volt regulator
1	Circuit board, approx. 2 × 3 inches
1	12-volt DC–AC adapter
4	10-µF, 16-volt capacitors
1	eight-position terminal strip
2	9-pin female "D" connectors
10 feet	two-conductor shielded wire
Miscellaneous	resistors and capacitors

Table 1-22: Parts list

levels, and therefore doesn't need special power supplies; and it's terminated with a resistor. These changes make the difference between a maximum of 100 feet and a maximum of 4,000 feet of wire that can be fed reliably. RS-232 can feed lengths of wire longer than 100 feet only if it is shielded.

The MAX232 produces the needed positive and negative 10 volts internally with the help of four external capacitors. If it doesn't work, check for the positive and negative 10 volts with a voltmeter. It takes the serial data and converts the level to TTL. Then the two 75176s per channel convert the TTL levels to differential RS-422 levels.

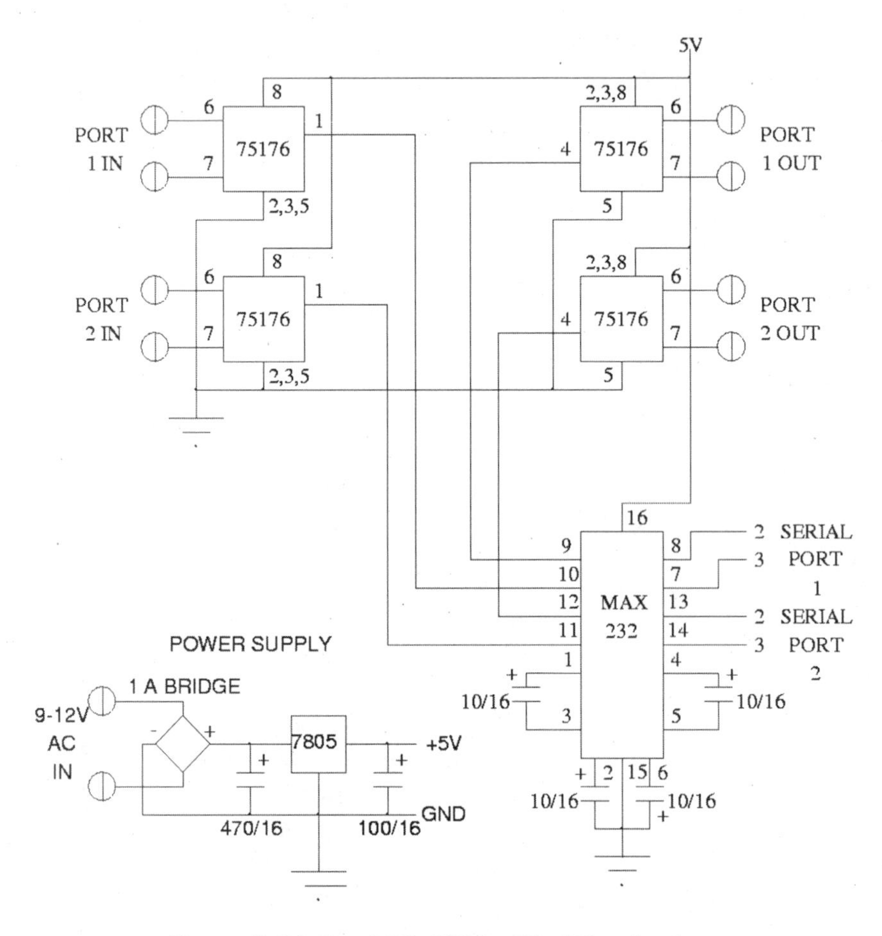

Figure 1-26: Dual RS-232 to RS-422 adapter

To test the converter, loop the device back to itself with two wires. At the RS-422 inputs, install a 100-ohm line-terminating resistor. Then bring up Windows Terminal and see if it talks to itself. It should echo every key press on the screen. With longer wire runs, two 470 ohm to 680 ohm resistors need to be added to the Port 1 and Port 2 inputs. One resistor connects to pin 6 and to +5 volts, the other connects to pin 7 and to ground.

This Quick Basic test program will select controller number 1, read its analog and digital input values, and cycle its relays from 1 to 8 sequentially. This can be used for the basis for a larger program to select multiple controllers and display their information. This information can then be written to a log file for viewing or charting and temperature.

```
'68 HC11 EMS TEST PROGRAM
OPEN "COM1: 9600,N,8,1,ASC,RS,CD0,CS0,DS0" FOR RANDOM AS #1
CYCLE = 0
START:
FOR A = 1 TO 300000: NEXT A    'Delay for Pentium computer
PRINT #1, CHR$(13);                       'Carriage return
PRINT #1, "1"                                     'Select device 1
PRINT #1, CHR$(2 ^ CYCLE);        'Cycle through 8 relays
CYCLE = CYCLE + 1
IF CYCLE = 8 THEN CYCLE = 0
INPUT #1, A$
PRINT "Received: ": A$
GOTO START
```

Chapter 2

Printer Port Projects

In this chapter you will discover:

Why use printer port devices? The answer is that they are available on old IBM PS-2s and on old laptops. We have had several old PS-2 model 55 386-SX based computers given to us. They used MCA for a proprietary expansion slot to stave off clones. The plan has backfired and led to piles of surplus computers that no one wants. Maybe the same thing will happen to Intel and its very similar approach to its Pentium II. Because PS-2 computers have no ISA slots, you cannot add internal devices to them. So using the printer port is the only way to add devices easily.

We also have some old laptop computers lying around. They have no PCMCIA slots for newer expansion devices. There, the only easy way of connecting external devices is once again via the parallel or serial ports.

Some old computers from Radio Shack have a proprietary printer port. This makes adding these devices difficult. You can buy or make an adapter to convert it to a normal 25-pin connector. You can also use an ISA Parallel port card to add a normal printer port to one, to get around the problem.

Let's take a look at the standard printer port. It has some oddities, with some signals inverted and some not connected in what would appear to be a logical order.

First, some pins are actually bidirectional. To use this feature you first place the output lines in a high state. Then the external device can send data in by pulling that output pin low. When I used this feature, I ran into BIOS incompatibilities. Some BIOSs would actually shut down the parallel port because they thought something was wrong with the port.

To get around this, wait until the computer is on to plug in the external device. Or, like me, choose not to use this bidirectional ability.

The bidirectional ability leads to another problem. In order to allow bidirectional compatibility, open collector driver ICs are used in the computer's printer port. Normal TTL drivers have both a transistor to pull the output high and one to pull it low. Open collector drivers, on the other hand, only pull the output low—there is nothing to pull it high.

To correct this condition, a "pull-up" resistor is needed on the control output lines, in every device that uses them. Some designs have specified values as high as 4.7K for this pull-up resistor. My experience has shown that a much lower value is needed, such as 220 ohm or 470 ohm. One end of the resistor connects to the control output pins 1, 14, 16, and 17, and the other connects to the 5-volt power supply. In extreme cases, a 0.01 or 0.001 capacitor from the control pin to ground will also help reduce noise pickup.

Other factors affecting noise pickup include the quality of the cable to the printer port, and even the computer's printer port card. I've been amazed at how these two factors affect the performance of both the printer port o'scope and the printer port video adapter.

Some of my BASIC programs will assume port 2. Port 2 is the most commonly available printer port. The programs run faster without the ability to select another port. If that port won't do for your application, then the source code will have to be changed to the port needed.

Pin	Bit	Direction	Inverted?	Signal Name
1.	D0	Control out	Yes	Data strobe
2.	D0	Data output		
3.	D1	Data output		
4.	D2	Data output		
5.	D3	Data output		
6.	D4	Data output		
7.	D5	Data output		
8.	D6	Data output		
9.	D7	Data output		
10.	D6	Control in		Acknowledge
11.	D7	Control in	Yes	Printer Busy
12.	D5	Control in		Printer empty
13.	D4	Control in		Select out
14.	D1	Control out	Yes	Auto feed
15.	D3	Control in	Yes	Interrupt
16.	D2	Control out		Initialize printer
17.	D3	Control out	Yes	Select in
18–25.	Ground			

Table 2-1: Printer port pins and specifications

Pin	Bit	Direction	Inverted?	Signal Name
2.	D0	Data output		
3.	D1	Data output		
4.	D2	Data output		
5.	D3	Data output		
6.	D4	Data output		
7.	D5	Data output		
8.	D6	Data output		
9.	D7	Data output		
15.	D3	Control in	Yes	Error
13.	D4	Control in		Select out
12.	D5	Control in		Printer empty
10.	D6	Control in		Acknowledge
11.	D7	Control in	Yes	Printer busy
1.	D0	Control out	Yes	Data strobe
14.	D1	Control out	Yes	Auto feed
16.	D2	Control out		Initialize printer
17.	D3	Control out	Yes	Select in
18-25.	Ground			

Table 2-2: Printer port pins sorted by groups

Port Number	Data Out	Control In	Control Out
1	3BC	3BD	3BE
2	378	379	37A
3	278	279	27A

Table 2-3: PRINTER I/O address assignments (change in program if needed)

Printer Port Video Adapter 2-1

The ability to see a video picture on your computer is fast becoming a standard. Not all of us, however, have the several hundred dollars necessary to buy the needed hardware and software. Some cheaper devices have come out to do the task, but they fall far short, taking several seconds for just one frame. But you can make an adapter yourself and do it for under $50! With a little experience in wire wrapping or printed circuit board assembly, including IC circuits, you can build yourself the printer port video adapter.

This adapter plugs into your computer's printer port. It grabs pictures at a resolution of 256×256 by 16 shades of gray format. The speed is one frame per one to two seconds on a 486 computer. Speed depends on computer speed and the resolution used. The only drawback is that the picture is in black and white.

Three Quick BASIC programs are included to view, save, and recall images. Both require a VGA monitor. One uses the 320 by 200 screen to allow up to 32 shades of gray. The image fills just about the whole monitor screen. The second program uses the 640 by 480 VGA screen mode. This allows separate live and recalled image areas, as well as a directory of saved images.

The third is like the second, but a whole lot slower. The speed is a trade-off for BMP compatibility.

In order to capture an image, you have to load it into some sort of memory. Then, you must be able to read that memory and show it on the screen or save it to a disk. The adapter contains its own memory of 64K by 8 bits. The adapter uses SRAM memory for high-speed operation. SRAM stands for "static random access memory." A 74LS157 multiplexer allows the adapter to select between two modes. The first is grabbing the picture into memory. The second is the computer reading the picture back out of memory. The computer then stores the picture into its memory, screen, and disk drives.

The adapter's memory is addressed by two 74LS393 counters. The first counter is the horizontal counter. It advances once for each pixel or dot from the left side to right side of the screen. The second, vertical counter advances once for each horizontal line, from the top to the bottom of the screen. The counters are also used to read data from the memory to the computer. In this case they are connected as though they are one sequential counter.

When capturing a picture, the counters are told when to count by synchronizing signals from an LM1881 sync separator. This IC takes a sample of the video input and separates the signals that tell how to synchronize the picture. Everything then lines up consistently. These signals are the horizontal and vertical sync signals.

A CA3306 high-speed video analog-to-digital converter takes the analog video pixels and converts them to digital values. These values are then stored in the SRAM memory.

A 74LS244 buffer sends the data back to the printer port of the computer 5 bits at a time. Only 4 bits are used, except in the lower resolution program. When the data gets to the computer, it has to be divided by 16 to return the bits to the correct position.

Another buffer, a 74LS14, receives control commands from the computer. It is also used as an inverter for the clock. It selects between the two memory chips and for D7 before it is sent back to the computer.

A 1N914 diode makes sure that the video level at sync time is 0 volts. This is referred to as "DC restoration." It prevents the average video level from wandering, which will happen with some VCRs.

An adjustable 5-MHz clock provides the master clock for the digital-to-analog converter. The oscillator has to be synchronized to the incoming signal. This is done by using a 4046 phase-locked loop. This PLL is synchronized so that the oscillator starts correctly with the left edge of the picture. Without this feature, vertical lines will have small steps in them. With it, every dot is synchronized with the left edge of the picture, and therefore, vertical lines are straight up and down.

The circuit can be built on a 4-1/2 × 6 inch circuit board, or on a printed circuit board. It fits into a box the same size (i.e., a little larger) by about 2 inches tall.

The printer port jack is a 25-pin male DIN type, to mate to the jack on the computer. The cable should be kept short, 4 to 6 feet maximum, to reduce noise pickup. The 220-ohm resistors, 0.001 capacitors, and 74LS14 make it possible to have that long of a cable.

Power comes from a regulated 5-volt 0.5-amp source. A 9-volt AC 500-mA adapter and an LM7805 regulator will work. The regulator will need a small heat sink.

The video input jack is a male RCA jack mounted on the box. An RCA cable then connects the input to the video output of a VCR, or to a camera. The video source must be strong or noise will be fed back into the source. A buffer could prevent this noise on weaker signals. On a VCR, this noise will appear as wavy lines through the picture of an attached TV. The noise only occurs while the video adapter is in the process of capturing a picture.

The only layout problem is with the analog parts. The CA3306 and LM1881 ICs are sensitive to and pick up digital noise. They must be kept isolated from the other circuitry and close to the input jack. A 470-ohm resistor and 500-pF capacitor at the input of the LM1881 help protect it from noise. Four or five capacitors about 0.1 μF in value from the 5-volt IC pins to their ground pins also reduce noise. Another noise reduction method is to use an 18- to 22-gauge copper wire to connect all of the grounds together. You could also use a two-layer board with one side at ground potential, as this forms a "bus" that is less prone to allow noise.

Mode selection via the 74LS157 is as follows:

```
COMPUTER IN CONTROL (D1=0)
________________________________
HORIZONTAL COUNT = COMPUTER COUNT - D0 = 1
HORIZONTAL RESET = COMPUTER RESET - D4 = 1
VERTICAL COUNT = HORIZONTAL COUNTER BIT 7
VERTICAL RESET = COMPUTER RESET
VIDEO IN CONTROL (D1=1)
____________________
HORIZONTAL COUNT = 5 MHz CLOCK
HORIZONTAL RESET = HORIZONTAL SYNC.
VERTICAL COUNT = HORIZONTAL COUNTER BIT 7
VERTICAL RESET = VERTICAL SYNC.
```

The printer port pin definitions chart, Table 2-4, shows how difficult it can be to input information via the printer port. The printer port is primarily designed as an output device, but it does have five inputs to monitor the printer status. These upper 5 bits arrive at the computer with 1 bit inverted. That bit has to be inverted before it is sent to the computer. Using software correction takes too much time.

The data bits then have to be divided by 16 when they get to the computer, to return the bits to the correct positions. This math is performed in the line containing "INP(&379)/16" in the Quick BASIC programs for high-resolution and BMP operation. In the low-resolution program it is all done, in advance, when the palette is defined. There is no software division necessary and the program runs faster. The two high-resolution programs utilize only the upper 4 bits, so they don't have to manipulate the data as much.

There are some areas of the picture that you don't want to capture. This is prevented by holding the appropriate counter at reset during those undesired times.

Data Bit	Inverted Data?	Input or Output?	Plug Pin#	Signal Name
Control output port:				
D0	Yes	O	1	Strobe data
D1	Yes	O	14	Auto feed paper
D2	No	O	16	Initialize printer
D3	Yes	O	17	Select in
Control input port:				
D3	Yes	I	15	Error
D4	No	I	13	Select out
D5	No	I	12	Printer empty
D6	No	I	10	Acknowledge
D7	Yes	I	11	Printer busy

Printer control I/O assignments:

Output data	Input data	LPT port number
3BE	3BD	1
37A	379	2
27A	279	3

Table 2-4: Printer port pin definitions

Bit	Value	Definition
Bit 0	Value of "1"	Next address
Bit 1	Value of "2"	Low = read, high = write
Bit 2	Value of "4"	Reset

Table 2-5: Control output assignments for video adapter

This "back porch" area refers to signals that appear just after the synchronization signals. The LM1881 doesn't have a vertical "back porch" detection circuit, so those signals are processed and stored into SRAM. Software is used to remove them. The vertical start position, used to draw the picture on the screen, begins as a negative number. This will result in starting above the top of the visible screen, which therefore hides those signals.

Back-porch signal applications include things such as closed captioning for the hearing impaired. During the horizontal back porch, at the left side of the screen, there is a color reference signal. This is a burst of green, but it sometimes appears as a white box along the side of the picture.

The software can run faster if the aspect ratio correction is omitted. This is the "IF X=5" line in the low-resolution program. Without it, the picture is a bit tall and narrow, but with it, the program runs a little more slowly. You can experiment with it by commenting out that line.

For price and availability, 32K × 8 SRAMS were used. These are used on motherboards as "cache memory." Speed is not a critical factor. At 5 MHz there are 100 nanoseconds for the counters to settle, then another 100 nanoseconds to write data into memory. Most of these cache chips can operate in the 15- to 25-nanosecond range. Size is a factor, and the cache SRAMS are quite small.

To get 64K of storage without taking the space and time for two IC sockets, I piggy-backed two 32K chips (i.e., put one on top of the other). Pin 20 on the

Quantity	Part Number	Source	Phone Number
1	LM1881	Digi-Key	1-800-344-4539
1	CA3306	Digi-Key	1-800-344-4539
2	32K X 8 SRAM	La-Paz	1-800-586-4199
1	74LS244	Mouser	1-800-346-6873
1	74LS158	Mouser	1-800-346-6873
2	74LS393	Mouser	1-800-346-6873
1	74HC4046	Mouser	1-800-346-6873
1	74LS14	Mouser	1-800-346-6873
1	1N914	Mouser	1-800-346-6873
1	28-pin narrow socket	Make from a wide one	
1	20-socket		
1	18-socket		
2	16-sockets		
3	14-sockets		
1	8-socket		
Misc.	resistors, capacitors, etc.	Radio Shack, Digi-Key, or Mouser	

Table 2-6: Parts sources

top chip is bent out first and connected to a short piece of wire that connects to pin 6 on the 74LS14. The two chips can then be soldered together. The run from pin 20 to ground is cut underneath the circuit board. A jumper under the circuit board connects the 74LS393 pin 3 to the 74LS14 pin 5, and to pin 20 of the SRAM socket as well.

To check out the operation of the device, first connect the 9-volt AC adapter. Then, plug in a short 25-pin to 25-pin cable to the printer port. Next, add an RCA cable to a VCR or other video output source. Make sure the VCR is on and in control, and that it has a good signal.

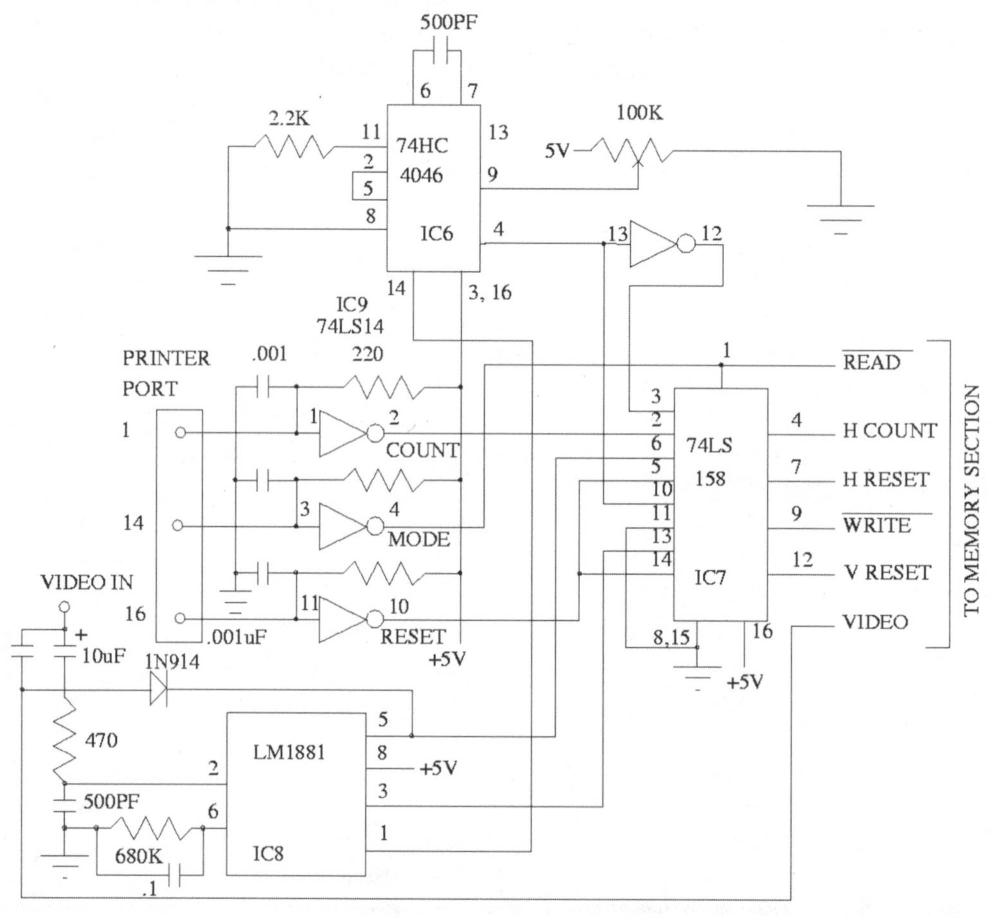

Figure 2-1: Printer port video control section

Use the software in Quick BASIC or the EXE version. When you run the software, there are three options. Pressing the letter "Q" quits the program. Pressing "S" saves the current picture from the video source. Pressing "V" lets you view a previously saved picture. It doesn't take much time to save a lot of pictures, and it's fun to be able to review them whenever you want.

There are three programs on the following pages. First there is the PPVA-FS, which is 320 × 200 with 32 shades of gray. Then there is PPVA-VGA, which is 640 × 480 with 16 shades of gray. The third is PPVA-BMP, which is 640 × 480 with BMP compatibility.

The BMP version has several lines that start with "DATA" to make a fake BMP file header. This header tells the program viewing the file the size of the picture, the size of the palette, and the actual colors to appear in the palette.

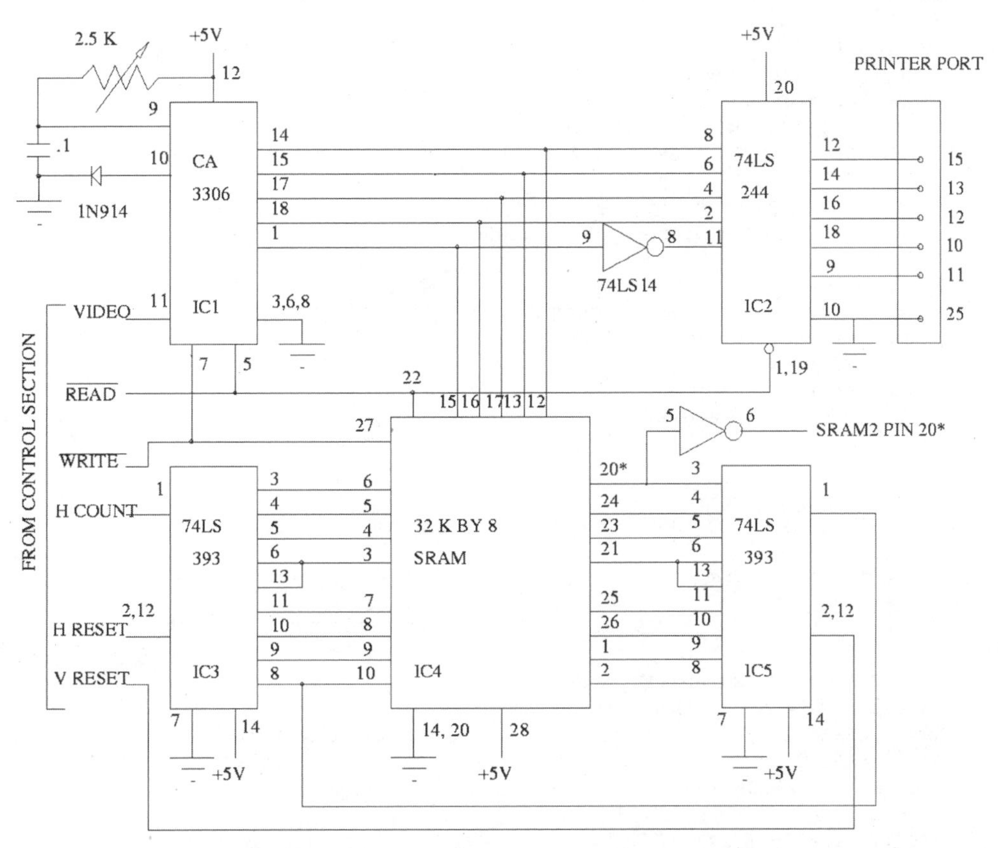

Figure 2-2: Printer port video memory section

```
'PRINTER PORT VIDEO ADAPTER - FULL SCREEN VERSION
'WRITTEN DECEMBER, 1994 BY BOB DAVIS
CLS : SCREEN 13                             '13 = 320 X 200 x 256
FOR I& = 0 TO 63                            'SET UP B&W PALETTE
PALETTE (I& * 4) + 7, (65536 * I& + 256 * I& + I&): NEXT I&
COLOR 127
LOCATE 1, 34: PRINT "PRINTER": LOCATE 2, 34: PRINT "PORT"
LOCATE 3, 34: PRINT "VIDEO": LOCATE 4, 34: PRINT "ADAPTER"
LOCATE 6, 34: PRINT "OPTIONS": LOCATE 7, 34: PRINT "Q=QUIT"
LOCATE 8, 34: PRINT "S=SAVE": LOCATE 9, 34: PRINT "V=VIEW"
START: OUT &H37A, 2                         'WRITE PICTURE
FOR A = 1 TO 10000: NEXT A                  'DELAY FOR V SYNC
OUT &H37A, 4                                'READ, RESET
FOR V = -12 TO 180                          'VERTICAL RANGE
X = X + 1: IF X = 5 THEN X = 1: V = V - 1'ASPECT RATIO FIX
PSET (0, V), 0
FOR H = 0 TO 255                         'HORIZONTAL RANGE
   OUT &H37A, 0: OUT &H37A, 1: PSET STEP(1, 0), INP(&H379)
NEXT H
NEXT V
key$ = INKEY$                               'GET USERS RESPONSE
IF key$ = "Q" OR key$ = "q" THEN END
DIM picture%(0 TO 29000)
IF key$ = "S" OR key$ = "s" THEN
LOCATE 25, 1: INPUT "NAME"; file$
GET (0, 0)-(252, 190), picture%
DEF SEG = VARSEG(picture%(1))
BSAVE file$, VARPTR(picture%(1)), 48000
DEF SEG
END IF
IF key$ = "V" OR key$ = "v" THEN
LOCATE 25, 1: INPUT "NAME"; file$
file$ = file$ + ".bas"
GET (0, 0)-(252, 190), picture%
DEF SEG = VARSEG(picture%(1))
BLOAD file$, VARPTR(picture%(1))
DEF SEG
PUT (0, 0), picture%, PSET
FOR A = 1 TO 100000: NEXT A              'DELAY FOR VIEWING
END IF
LOCATE 15, 34: PRINT "        ": LOCATE 16, 34: PRINT "        "
GOTO START                                  'RESTART
```

```
'PRINTER PORT VIDEO ADAPTER 640 X 480 RES
'WRITTEN, 1994 BY ROBERT DAVIS
CLS : SCREEN 12                             '12 = 640 x 480 x 16
FOR I& = 0 TO 15                            'SET UP B&W PALETTE
PALETTE I&, (65536 * I& + 256 * I& + I&) * 4: NEXT I&
COLOR 10
LINE (1, 1)-(256, 247), , B
```

```
LINE (3, 3)-(254, 245), , B
LINE (300, 1)-(556, 247), , B
LINE (303, 3)-(554, 245), , B
LOCATE 16, 10: PRINT "LIVE VIDEO"
LOCATE 16, 48: PRINT "STORED PICTURE"
LOCATE 17, 1: PRINT "** PRINTER PORT VIDEO ADAPTER **"
LOCATE 17, 40: PRINT "OPTIONS; Q=QUIT, S=SAVE, V=VIEW"
LOCATE 18, 1: PRINT "AVAILABLE FILES;"
LOCATE 19, 1: FILES "*.MPX"
START: OUT &H37A, 2                      'WRITE PICTURE DATA TO RAM
FOR A = 1 TO 5000: NEXT A                'DELAY FOR V SYNC
OUT &H37A, 4                             'READ DATA, RESET
FOR V = -16 TO 240                       'VERTICAL RANGE
PSET (0, V), 0
FOR H = 0 TO 255                     'HORIZONTAL RANGE
  OUT &H37A, 0: OUT &H37A, 1: PSET STEP(1, 0), INP(&H379) / 16
NEXT H
NEXT V
key$ = INKEY$                            'GET USERS RESPONSE
IF key$ = "Q" OR key$ = "q" THEN END
DIM PICTURE%(0 TO 29000)      ' WON'T COMPILE IF GREATER THAN 29K
IF key$ = "S" OR key$ = "s" THEN
LOCATE 18, 40: INPUT "SAVE FILE NAME? ", file$
LOCATE 18, 40: PRINT "                              "
file$ = file$ + ".MPX"
GET (0, 0)-(255, 240), PICTURE%
DEF SEG = VARSEG(PICTURE%(1))
BSAVE file$, VARPTR(PICTURE%(1)), 32000
DEF SEG
END IF
IF key$ = "V" OR key$ = "v" THEN
LOCATE 18, 40: INPUT "VIEW FILE NAME? ", file$
LOCATE 18, 40: PRINT "                              "
file$ = file$ + ".MPX"
GET (0, 0)-(255, 240), PICTURE%
DEF SEG = VARSEG(PICTURE%(1))
BLOAD file$, VARPTR(PICTURE%(1))
DEF SEG
PUT (300, 0), PICTURE%, PSET
END IF
LOCATE 18, 1: PRINT "AVAILABLE FILES;"
LOCATE 19, 1: FILES "*.MPX"
GOTO START                        'RESTART FOR CONTINUOUS UPDATING
```

```
'PRINTER PORT VIDEO ADAPTER - BMP VERSION
'BMP MOD WRITTEN FEBUARY, 1996 BY BOB DAVIS
CLS : SCREEN 12                          '12 = 640 x 480 x 16
FOR I& = 0 TO 15:PALETTE I&, (65536 * I& + 256 * I& + I&) * 4
NEXT I&
COLOR 10
```

```
LINE (1, 1)-(256, 247), , B: LINE (3, 3)-(254, 245), , B
LINE (300, 1)-(555, 247), , B: LINE (303, 3)-(553, 245), , B
LOCATE 16, 10: PRINT "LIVE VIDEO"
LOCATE 16, 48: PRINT "STORED PICTURE"
LOCATE 17, 1: PRINT "** PRINTER PORT VIDEO ADAPTER **"
LOCATE 17, 40: PRINT "OPTIONS; Q=QUIT, S=SAVE, V=VIEW"
LOCATE 18, 1:PRINT "AVAILABLE FILES;":LOCATE 19, 1: FILES "*.BMP"
START: OUT &H37A, 2                      'WRITE PICTURE DATA TO RAM
FOR A = 1 TO 6000: NEXT A                'DELAY FOR V SYNC
OUT &H37A, 4                             'READ DATA, RESET
DIM BARY%(0 TO 29600): V2 = -16         'STORE REVERSE & WRITE NORMAL
FOR V = 250 TO 0 STEP -1                 'STORE IN ARRAY & ON SCREEN
 B = V * 128
 PSET (0, V2), 0: V2 = V2 + 1
 FOR H = 0 TO 127
 OUT &H37A, 0: OUT &H37A, 1: C = INP(&H379) AND &HF0: PSET STEP
   (1, 0), C / 16
 OUT &H37A, 0:OUT &H37A, 1:D = INP(&H379) \ 16:PSET STEP(1, 0), D
  IF V <= 230 THEN BARY%(B) = C + D: B = B + 1
 NEXT H
 NEXT V
key$ = INKEY$                              'GET USERS RESPONSE
IF key$ = "Q" OR key$ = "q" THEN END
IF key$ = "S" OR key$ = "s" THEN         'SAVE PICTURE TO DISK
LOCATE 18, 40: INPUT "SAVE FILE NAME? ", FILE$
LOCATE 18, 40: PRINT "                                   "
FILE$ = FILE$ + ".BMP": OPEN FILE$ FOR BINARY AS #1
DATA 66,77,246,115,0,0,0,0,0,0,118,0,0,0,40,0
DATA 0,0,255,0,0,0,230,0,0,0,1,0,4,0,0,0
DATA 0,0,0,0,0,0,0,0,0,0,0,0,0,0,0,0,0,0,0,0,0,0
DATA 0,0,0,0,16,16,16,0,32,32,32,0,48,48,48,0
DATA 64,64,64,0,80,80,80,0,96,96,96,0,112,112,112,0
DATA 128,128,128,0,144,144,144,0,160,160,160,0,176,176,176,0
DATA 192,192,192,0,208,208,208,0,224,224,224,0,240,240,240,0
FOR A = 1 TO 118: READ B: C$ = CHR$(B): PUT #1, A, C$: NEXT A
FOR A = 119 TO 29559: PUT #1, (A), BARY%(A - 119): NEXT A
RESTORE: CLOSE #1
END IF
IF key$ = "V" OR key$ = "v" THEN         'GET PICTURE FROM DISK
LOCATE 18, 40: INPUT "VIEW FILE NAME? ", FILE$
LOCATE 18, 40: PRINT "                                    "
FILE$ = FILE$ + ".BMP": OPEN FILE$ FOR BINARY AS #1
B = 1
FOR V = 230 TO 1 STEP -1             'REVERSE ORDER FOR .BPB FILE
 PSET (300, V), 0
 FOR H = 1 TO 128
```

```
    GET #1, B + 118, BARY%(B)
    PSET STEP(1, 0), (BARY%(B) AND &HF0) / 16
    PSET STEP(1, 0), BARY%(B) AND &HF
    B = B + 1
  NEXT H
NEXT V: CLOSE #1
END IF
LOCATE 18, 1: PRINT "AVAILABLE FILES;":LOCATE 19, 1:FILES "*.BMP"
GOTO START                                    'RESTART FOR CONTINUOUS UPDATING
```

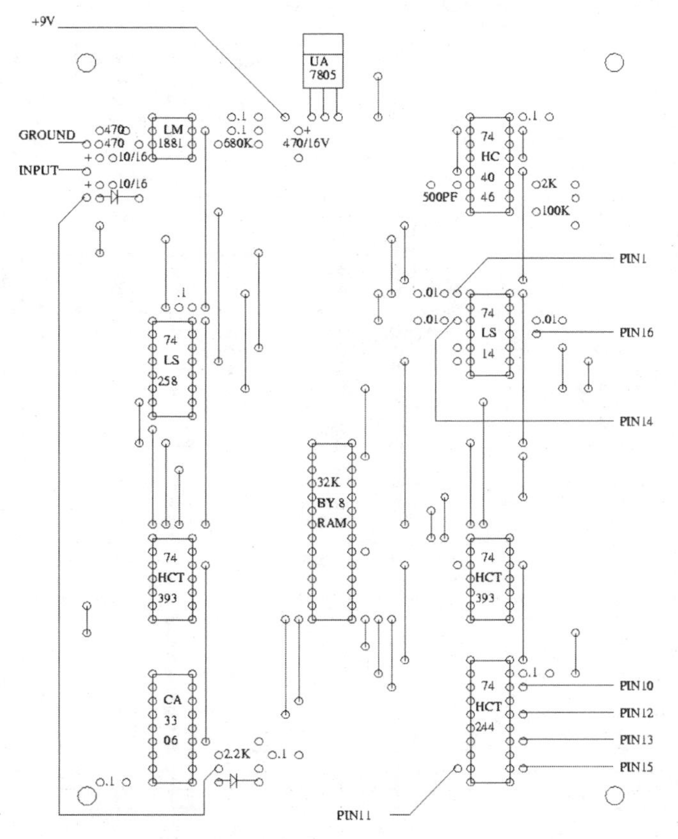

Figure 2-3: Printer port video parts layout

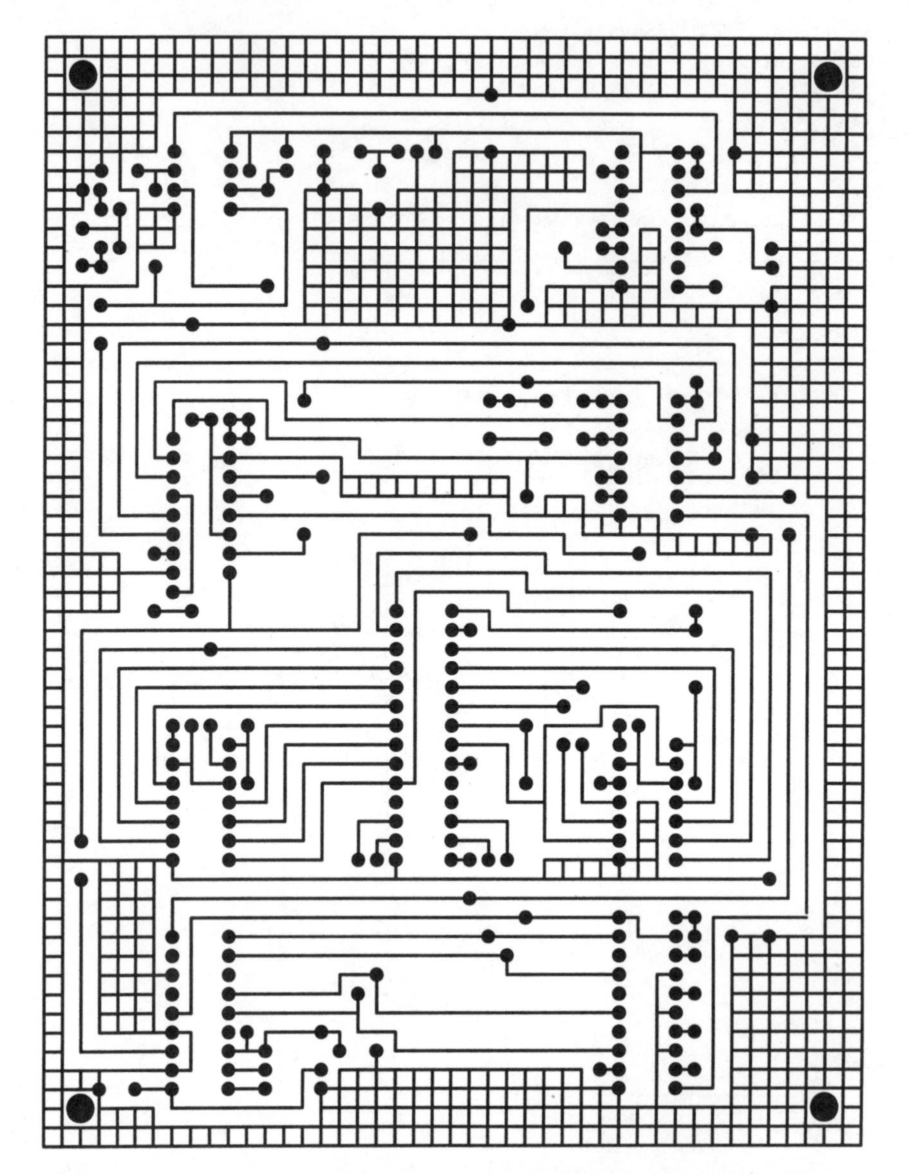

Figure 2-4: Printer port video circuit board design (top view)

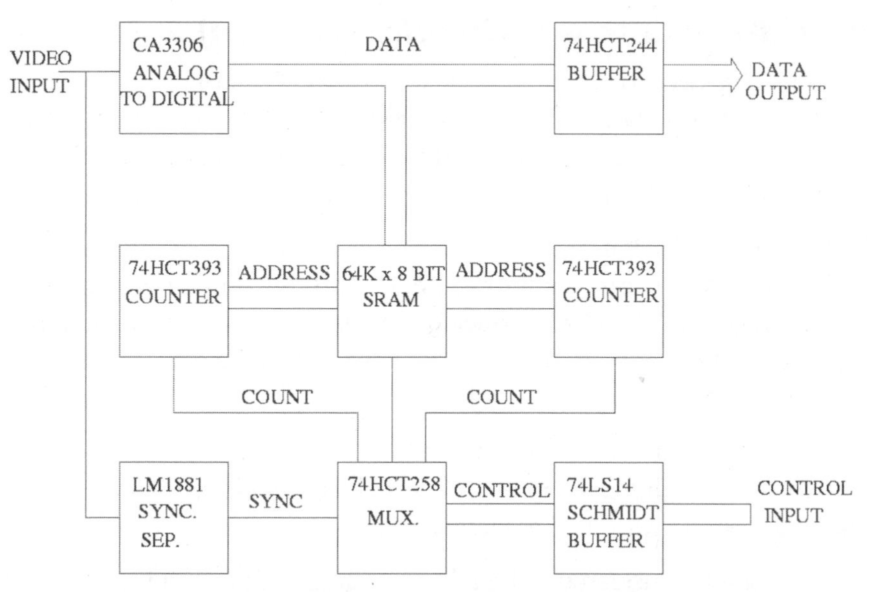

Figure 2-5: Printer port video block diagram

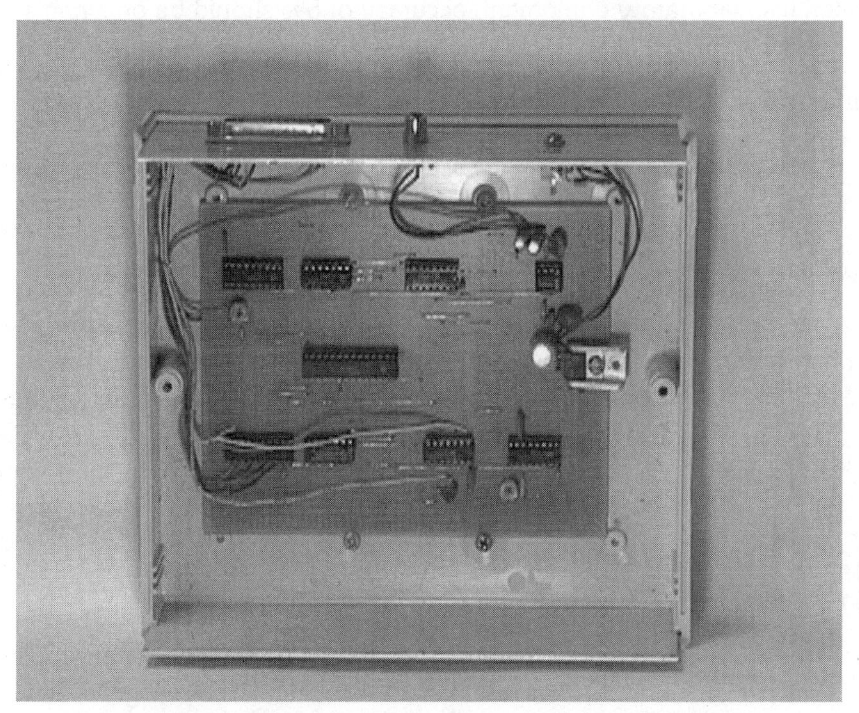

Figure 2-6: Picture of the printer port video adapter

Printer Port O'scope (Simpler Version) 2-2

Hand-held digital oscilloscopes are quite expensive. They often sell for more than $1,000. By using your laptop or desktop computer, you can slash that price substantially. In fact, you can build a single-channel version for under $100. Any old laptop with only a floppy drive can run the software.

This is a simplified version of the printer port oscilloscope. The fancier dual-channel version is too complex for beginners, and this simpler version can be built on a printed circuit board, reducing the chances of errors. It could be made into a dual-channel version, but both channels would share the same sensitivity settings.

This version is also missing the trigger section. Instead, the computer grabs samples and looks through the samples for a signal that would have triggered it. It then displays starting with the "triggering" signal.

This oscilloscope is considered to be a "hobby-grade" device. It is not suitable for medical or professional applications. Design accuracy is 10%, although if it were calibrated with laboratory equipment, accuracy of 5% should be obtainable.

The printer port oscilloscope starts with an input buffer and voltage divider feeding an amplifier. This then drives the analog-to-digital converter and feeds a FIFO (first-in first-out) memory buffer. The output of the FIFO is then multiplexed into two 4-bit chunks that can be input into the printer port.

A clock is derived from a series of clock dividers and selected by eight-to-one multiplexers.

The key components are a high-speed analog-to-digital converter and a high-speed memory device. Thanks to a new chip from Analog Devices, the AD775, the first problem is resolved for $15. This compares to $35 or more for the old 8-bit, 15-MHz analog-to-digital converter, the CA3318. The AD775 is available from Newark Electronics. Because of high demand, it might be in short supply.

There are four problems to deal with using the AD775. The first problem is that the AD775 is 0.4 inches wide. The required socket can be made from a normal 24-pin socket. First, however, you must cut a strip out of the center.

Second, the AD775 is rated for 20-MHz guaranteed operation. For a sampling rate of 2 μs per division, 25-MHz operation is needed. The Analog Devices specification sheet shows that the AD775 can work up to 35 MHz, with a slight loss of accuracy.

The third problem is that a reference supply is needed. For professional applications, two high-precision references can be used. For our application, a handpicked 78L05 or 78M05 will suffice. It needs to be within 4.9 to 5.1 volts to be accurate enough.

Last, the 5-volt reference leads to another problem. In order to sample signals below 0 volts, they must be biased so that they are above 0 volts. The AD775 has a low end of 0.6 volts and a high end of 2.7 volts on the internal resistor divider. The AD775 would then need an average input voltage of 1.7 volts. This is done with a NE592 video amplifier. It naturally has a bias on its output of about 2.5 volts, depending on its gain. A 2N3904 transistor then steps that down to 1.9V.

To keep the chip count to a minimum, I used a FIFO memory IC instead of SRAM. This FIFO replaces two counters, multiplexers, and an SRAM memory chip. FIFOs used to be quite expensive, but now are available for around $10 to $25, depending on the amount of memory and speed of the device. "First-in first-out." means that, information is put into the FIFO and retrieved from it in the same order. This is much like a shift register, except that the FIFO is 9 bits wide. FIFOs also feature "dual ports."

Dual ports allow information to be read from and written into the FIFO at the same time and at two different locations in memory. The ability to read and write at the same time is not being used in this application. In order to keep digital noise to a minimum, we will not read from the FIFO while in the write mode.

FIFOs require special handshaking to prevent them from going into "flow-through mode" where they essentially disappear from the circuit. If a read is attempted when the FIFO is empty or a write is attempted when the FIFO is full, it enters this "flow-through mode." After that, every read or write flows directly from input to output.

To prevent "flow-through," a four-input gate pulls the write high whenever there is a full flag or a reset. Software prevents reading when the FIFO is empty. But once you start reading from the FIFO, it starts writing again because it is no longer "full." Write inputs at this time will be full of noise signals since the read buffers are sending data. So be careful not to read from more than 2048 addresses at a time unless the FIFO is reset. That is based on using a 2K FIFO; actually, a 512-byte FIFO would suffice. This would, however, limit the capture time to little more than what you can see on the screen with the horizontal position at "0."

When it comes to speed, the Dallas DS2010 FIFO is officially rated at 50 ns. For proper operation at 25 MHz a speed in excess of 40 ns is needed. Every Dallas chip I have tested worked at 25 MHz, but that is no guarantee of proper operation. The AM7203A-25 is twice as expensive, but is rated to work at speeds in excess of 25 MHz.

Two ICs are used to determine the mode, a 74HCT20 and a 74HCT14. The 74LS20 makes sure that noise doesn't trigger a reset. It also stops write inputs once the FIFO is full. A 74LS257 uses the four control input lines to send data, 4 bits at a time, back to the computer. A 74LS14 buffers control lines and inverts the output data lines to provide true data. The 74HCT14 is a Schmidt trigger used to reduce noise and an inverter used to correct for some signals being inverted from the printer port. The state of D1 makes it possible to select between two different channels.

The next problem is that the clock speed of 25 MHz is fine for high-speed mode, but it needs to be slowed down in order to capture lower frequencies. This has to be done in an orderly manner so you can know exactly what the frequency is, and hence what the sample rate is. Usually a 1–2–5–10 stepping sequence method is used. The 74LS390 contains two of the dividers that are needed in each chip. Each divides by 2 and then 5, for a total of 10.

To divide 25 MHz down to 2,500 cycles, two of these ICs are needed. This gives a total of four divide-by-10s. Then two 74LS251 selectors select the desired clock frequency out of the 16 possibilities. To choose between the two selectors, data bits 3 and 4 are used. The selected clock output is also called the write clock or system clock.

FETs provide high impedance in the input and attenuation circuits. This is necessary to minimize loading on the circuit under test. The FETs don't, however, provide gain. A 2N3904 provides low-impedance drive to the resistor voltage divider network.

The power supply limits the entire input section to 12 volts peak-to-peak signal. By dividing the signal by 10 using a 10× attenuation probe, inputs of up to 120 volts can be displayed on the oscilloscope. Trimmer capacitors on the input dividers are needed to allow adjustment of the frequency response of the attenuator. This is adjusted so that a square wave will be properly displayed.

Then there is the input attenuation circuit. Using data from the port, a 4051 analog multiplexer selects a tap on a resistor divider to feed the NE592 amplifier. This divider allows a range of 0.02 to 2 volts per division. The NE592 amplifier provides inversion, adjustable gain, and a bias voltage of 1.7 volts. The gain needs to be adjusted so that a 1-volt peak-to-peak signal at 0.1 volts per division fills the entire screen. Also, a 0.5-volt peak-to-peak signal could be used to fill five divisions, or one-half of the screen.

The input voltage divider uses some unusual resistor values. You can get special high-precision resistors, or you can make them yourself. The 600-ohm resistor is made up of two 1.2-K resistors in parallel. The 60-ohm resistor consists of two 120-ohm resistors in parallel. The 200-ohm resistor comprises two 100-ohm resistors in series. The 20-ohm resistor is made up of two 10-ohm resistors connected in series.

The trigger circuit depends totally on software. The assumption is that the desired trigger level can be captured in the 2,048 samples taken during a scan. The wave form is then displayed on the computer from the trigger point on.

The circuit is built on a pad per hole circuit board, or a printed circuit board. On a pad per hole board power, ground, and analog ground are created using solid

Position	Frequency	Time
13	25 MHz	2 µs per division
12	10 MHz	5 µs per division
11	5 MHz	10 µs per division
10	2.5 MHz	20 µs per division
9	1 MHz	50 µs per division
8	500 kHz	100 µs per division
7	250 kHz	200 µs per division
6	100 KHz	500 µs per division
5	50 kHz	1 ms per division
4	25 kHz	2 ms per division
3	10 kHz	5 ms per division
2	5 kHz	10 ms per division
1	2.5 kHz	20 ms per division

Table 2-7: Frequency to time conversion

Quantity	Part Number	Source
1	NE592	Newark
1	4051	Newark
1	AD775	Newark
1	DS2010	Newark
1	74LS14	Newark
1	74LS257	Newark
1	74LS20	Newark
1	25 MHz clock	Newark
2	74LS151	Newark
2	73LS390	Newark
8	14-pin IC sockets	Newark
2	16-pin IC sockets	Newark
1	24-pin IC socket	Newark
1	28-pin IC socket	Newark
2	BNC connectors	Newark
1	circuit board, 4.5 inches × 6 inches	Newark
2	2N4416 FET transistors	Newark
2	2N3904 transistors	Newark
1	LM78L05 regulator	Newark
1	LM79L05 regulator	Newark
1	25-pin male connector	Newark
4	1N4148 diodes	Newark
1	cabinet approx. 10 inches × 8 inches × 3 inches	Mouser
1	12.6 VCT 1.2A transformer	Radio Shack
1	bridge rectifier, 1 amp, 50 V	Radio Shack
1	power cord and switch	Radio Shack
Miscellaneous	resistors and capacitors	Radio Shack

Table 2-8: Materials list for printer port oscilliscope (simpler version)

copper wire. After the power distribution is connected, check it out with a voltmeter for the proper voltages. Then it is safe to install the ICs.

To make the printed circuit board, the design is photocopied onto one of the many circuit board transfer films. Then the design is transferred from the film to a copper-clad board with a household iron. The film is removed, and the board is etched in acid. Once etched, the toner is removed with sandpaper. Holes are

Bits	Binary	Mode Selected
D0	1	Selects high or low 1/2 byte
D1	2	Select channel 1 or channel 2
D2	4	None
D3	8	Selects read or write data
D0-D3	15	Reset the scope

Table 2-9: Control line mode selection chart

Data Out	Control Out	Control In	Port Number
3BC	3BE	3BD	1
378	37A	379	2
278	27A	279	3

Table 2-10: Printer I/O assignments (change in program if needed)

Volts/division	3 bits gain (D5–7)
Time/division	3 bits (D0–2), 2 select bits (D3, D4)

Table 2-11: Controls and how they work

then drilled for the component leads. Harbor Freight has a drill press for $50, and the bits can be picked up at ham festivals for as little as $3 a set.

If you are not using the printed circuit board, the NE592s and 4051s cannot be wire wrapped. You must instead use point-to-point soldering, with analog ground buses surrounding their area on the circuit board.

The wiring can be done one section at a time. Each section is then wired and tested with a logic probe or an o'scope for proper operation. The AD775 can be tested with a direct input as long as its voltage is between ground and 5 volts.

To reduce noise at the inputs, which might be induced into the wire running to the printer port, 220-ohm to 470-ohm resistors connected to the 5-volt supply can be used. These input-pins are 1 to 9 and 14 to 17. Pins 20 through 25 are tied to the ground.

There are two software programs that will make this work. The first program is PSCOPE. The second is PSCOPEV, a higher resolution version.

```
' PRINTER PORT OSCILLOSCOPE II 10/01/95 BY BOB DAVIS
' LOW RESOLUTION VERSION FOR 640 X 200 CGA
CLS : LOCATE 10, 20: INPUT "USE PRINTER PORT NUMBER: ", LPT
DOUT = &H378: COUT = &H37A: CIN = &H379
IF LPT = 1 THEN DOUT = &H3BC: COUT = &H3BE: CIN = &H3BD
IF LPT = 3 THEN DOUT = &H278: COUT = &H27A: CIN = &H279
CLS : SCREEN 2
LINE (0, 0)-(639, 190), , B
LINE (0, 0)-(532, 190), , B
LINE (536, 4)-(635, 186), , B
LINE (561, 62)-(610, 72), , B
LINE (561, 94)-(610, 104), , B
LINE (561, 126)-(610, 136), , B
LOCATE 2, 71: PRINT "PRINTER"
LOCATE 3, 71: PRINT " PORT "
LOCATE 4, 71: PRINT "O'SCOPE"
LOCATE 7, 69: PRINT "CH 1 V/DIV."
LOCATE 9, 72: PRINT "  1"
LOCATE 11, 69: PRINT "CH 1 POSITN"
LOCATE 13, 72: PRINT "   0"
LOCATE 15, 69: PRINT "HORIZ T/DIV"
LOCATE 17, 72: PRINT " 5 mS"
LOCATE 21, 70: PRINT "COPYRIGHT"
LOCATE 22, 70: PRINT "BOB DAVIS"
LOCATE 25, 15
PRINT "SELECT CONTROLS WITH NUMERIC ARROW KEYS OR Q=QUIT";
T = 13: S = 2: P = 0: C = 9
START: OUT COUT, 11
OUT DOUT, (T + 32 * S)                              'CONTROL OUT
OUT COUT, 15                                         'RESET - START
FOR A = 1 TO 1000: NEXT A                           'CAPTURE DELAY
FOR H = 1 TO 530                                     'HORIZONTAL RANGE
PSET (H, 2), 0: DRAW "C0" + "M=" + VARPTR$(H) + ",189"
IF H > 2 THEN PSET (H - 1, V), 0 ELSE PSET (H, 95), 0
OUT COUT, 4: V = (INP(CIN) \ 16)       'GET LOW 1/2 BYTE
OUT COUT, 5: V = V + (INP(CIN) AND &HF0) - P - 30
IF V > 188 THEN V = 188 ELSE IF V < 2 THEN V = 2
IF H > 2 THEN DRAW "C1" + "M=" + VARPTR$(H) + ",=" + VARPTR$(V)
OUT COUT, 0: OUT COUT, 8                   'NEXT ADDRESS
NEXT H
key$ = INKEY$: IF key$ = "q" OR key$ = "Q" THEN END
LOCATE C, 69: PRINT "  "
IF key$ = "2" THEN C = C + 4: IF C = 21 THEN C = 9
IF key$ = "8" THEN C = C - 4: IF C = 5 THEN C = 17
LOCATE C, 69: PRINT "->"
IF C = 9 THEN                                        'SENSITIVITY
IF key$ = "6" THEN : S = S + 1: IF S > 7 THEN S = 0
```

```
IF key$ = "4" THEN : S = S - 1: IF S < 0 THEN S = 7
IF S = 0 THEN LOCATE 9, 72: PRINT "GND"
IF S = 1 THEN LOCATE 9, 72: PRINT "  2"
IF S = 2 THEN LOCATE 9, 72: PRINT "  1"
IF S = 3 THEN LOCATE 9, 72: PRINT " .5"
IF S = 4 THEN LOCATE 9, 72: PRINT " .2"
IF S = 5 THEN LOCATE 9, 72: PRINT " .1"
IF S = 6 THEN LOCATE 9, 72: PRINT ".05"
IF S = 7 THEN LOCATE 9, 72: PRINT ".02"
END IF
IF C = 13 THEN                                  'POSITION
IF key$ = "6" THEN : P = P + 2: IF P > 30 THEN P = -30
IF key$ = "4" THEN : P = P - 2: IF P < -30 THEN P = 30
LOCATE 13, 73: PRINT P
END IF
IF C = 17 THEN                                 'TIME BASE
IF key$ = "6" THEN : T = T + 1: IF T > 23 THEN T = 11
IF key$ = "4" THEN : T = T - 1: IF T < 11 THEN T = 23
IF T = 11 THEN LOCATE 17, 72: PRINT "20 mS"
IF T = 12 THEN LOCATE 17, 72: PRINT "10 mS"
IF T = 13 THEN LOCATE 17, 72: PRINT " 5 mS"
IF T = 14 THEN LOCATE 17, 72: PRINT " 2 mS"
IF T = 15 THEN LOCATE 17, 72: PRINT " 1 mS"
IF T = 16 THEN LOCATE 17, 72: PRINT ".5 mS"
IF T = 17 THEN LOCATE 17, 72: PRINT ".2 mS"
IF T = 18 THEN LOCATE 17, 72: PRINT ".1 mS"
IF T = 19 THEN LOCATE 17, 72: PRINT "50 uS"
IF T = 20 THEN LOCATE 17, 72: PRINT "20 uS"
IF T = 21 THEN LOCATE 17, 72: PRINT "10 uS"
IF T = 22 THEN LOCATE 17, 72: PRINT " 5 uS"
IF T = 23 THEN LOCATE 17, 72: PRINT " 2 uS"
END IF
GOTO START
```

```
' PRINTER PORT OSCILLOSCOPE II 10/01/95 BY BOB DAVIS
' HIGH RESOLUTION VERSION FOR 640 X 400 VGA
CLS : LOCATE 10, 20: INPUT "USE PRINTER PORT NUMBER: ", LPT
DOUT = &H378: COUT = &H37A: CIN = &H379
IF LPT = 1 THEN DOUT = &H3BC: COUT = &H3BE: CIN = &H3BD
IF LPT = 3 THEN DOUT = &H278: COUT = &H27A: CIN = &H279
CLS : SCREEN 9
LINE (550, 0)-(639, 329), 1, BF
LINE (572, 94)-(617, 114), 0, BF
LINE (572, 150)-(617, 170), 0, BF
LINE (572, 206)-(617, 226), 0, BF
LINE (572, 262)-(617, 282), 0, BF
LINE (0, 0)-(639, 329), , B
LINE (550, 0)-(639, 329), , B
LINE (554, 4)-(635, 325), , B
LINE (572, 94)-(617, 114), , B
```

```
LINE (572, 150)-(617, 170), , B
LINE (572, 206)-(617, 226), , B
LINE (572, 262)-(617, 282), , B
LOCATE 2, 72: PRINT "PRINTER"
LOCATE 3, 72: PRINT " PORT  "
LOCATE 4, 72: PRINT "O'SCOPE"
LOCATE 6, 71: PRINT "VOLTS/DIV"
LOCATE 10, 71: PRINT "POSITION "
LOCATE 14, 71: PRINT "TIME/DIV "
LOCATE 18, 71: PRINT " TRIGGER "
LOCATE 22, 71: PRINT "COPYRIGHT"
LOCATE 23, 71: PRINT "BOB DAVIS"
LOCATE 8, 73: PRINT " 1.0 "
LOCATE 12, 73: PRINT "  0  "
LOCATE 16, 73: PRINT "  5MS"
LOCATE 20, 73: PRINT "  5  "
LOCATE 25, 15: PRINT "SELECT CONTROLS WITH NUMERIC ARROW KEYS OR Q=QUIT";
T = 13: S = 2: P = 0: C = 8: R = 5: V = 128
START: OUT COUT, 11
OUT DOUT, (T + 32 * S): OUT COUT, 15    'RESET - START
FOR A = 1 TO 1000: NEXT A                'CAPTURE DELAY
CIRCLE (625, 272), 5, 7
PAINT (625, 272), 0, 7
MTRIG = 0: TRIGD = 0: RA = R + 128
TLOOP:                                    'TRIGGER CONTROL ROUTINE
OUT COUT, 4: V = (INP(CIN) \ 16)
OUT COUT, 5: V = V + (INP(CIN) AND &HF0)
OUT COUT, 0: OUT COUT, 8             'NEXT ADDRESS
MTRIG = MTRIG + 1
IF MTRIG < 999 THEN
IF TRIGD = 0 THEN
  IF (V > RA AND RA < 0) OR (V < RA AND RA > 0) THEN TRIGD = 1
   GOTO TLOOP
END IF
IF (V < RA AND R < 0) OR (V > RA AND R > 0) THEN
   PAINT (625, 272), 4, 7
ELSE
   GOTO TLOOP
END IF
END IF
FOR H = 1 TO 549                          'HORIZONTAL RANGE
OUT COUT, 4: V = (INP(CIN) \ 16)     'GET LOW 1/2 BYTE
OUT COUT, 5: V = V + (INP(CIN) AND &HF0) - P + 50
IF V > 327 THEN V = 327 ELSE IF V < 2 THEN V = 2
IF H > 2 THEN LINE (H, 1)-(H, 328), 0
IF H > 2 THEN LINE (H, OLDV)-(H, V)
FOR VL = 30 TO 300 STEP 30: PSET (H, VL), 2: NEXT VL
IF H MOD 50 = 0 THEN LINE (H, 1)-(H, 328), 2
OLDV = V: OUT COUT, 0: OUT COUT, 8  'NEXT ADDRESS
NEXT H
key$ = INKEY$: IF key$ = "q" OR key$ = "Q" THEN END
```

```
LOCATE C, 73: PRINT " "
IF key$ = "2" THEN C = C + 4: IF C = 24 THEN C = 8
IF key$ = "8" THEN C = C - 4: IF C = 4 THEN C = 20
LOCATE C, 73: PRINT ">"
IF C = 8 THEN                           'SENSITIVITY
IF key$ = "6" THEN : S = S + 1: IF S > 7 THEN S = 0
IF key$ = "4" THEN : S = S - 1: IF S < 0 THEN S = 7
IF S = 0 THEN LOCATE 8, 74: PRINT "GND "
IF S = 1 THEN LOCATE 8, 74: PRINT "2.0 "
IF S = 2 THEN LOCATE 8, 74: PRINT "1.0 "
IF S = 3 THEN LOCATE 8, 74: PRINT " .5 "
IF S = 4 THEN LOCATE 8, 74: PRINT " .2 "
IF S = 5 THEN LOCATE 8, 74: PRINT " .1 "
IF S = 6 THEN LOCATE 8, 74: PRINT " .05"
IF S = 7 THEN LOCATE 8, 74: PRINT " .02"
END IF
IF C = 12 THEN                          'POSITION
IF key$ = "6" THEN : P = P + 2: IF P > 90 THEN P = -90
IF key$ = "4" THEN : P = P - 2: IF P < -90 THEN P = 90
LOCATE 12, 74: PRINT P
END IF
IF C = 16 THEN                          'TIME BASE
IF key$ = "6" THEN : T = T + 1: IF T > 23 THEN T = 11
IF key$ = "4" THEN : T = T - 1: IF T < 11 THEN T = 23
IF T = 11 THEN LOCATE 16, 74: PRINT "20MS"
IF T = 12 THEN LOCATE 16, 74: PRINT "10MS"
IF T = 13 THEN LOCATE 16, 74: PRINT " 5MS"
IF T = 14 THEN LOCATE 16, 74: PRINT " 2MS"
IF T = 15 THEN LOCATE 16, 74: PRINT " 1MS"
IF T = 16 THEN LOCATE 16, 74: PRINT ".5MS"
IF T = 17 THEN LOCATE 16, 74: PRINT ".2MS"
IF T = 18 THEN LOCATE 16, 74: PRINT ".1MS"
IF T = 19 THEN LOCATE 16, 74: PRINT "50US"
IF T = 20 THEN LOCATE 16, 74: PRINT "20US"
IF T = 21 THEN LOCATE 16, 74: PRINT "10US"
IF T = 22 THEN LOCATE 16, 74: PRINT " 5US"
IF T = 23 THEN LOCATE 16, 74: PRINT " 2US"
END IF
IF C = 20 THEN                          'TRIGGER
IF key$ = "6" THEN : R = R + 1: IF R > 30 THEN R = -30
IF key$ = "4" THEN : R = R - 1: IF R < -30 THEN R = 30
LOCATE 20, 74: PRINT R
END IF
GOTO START
```

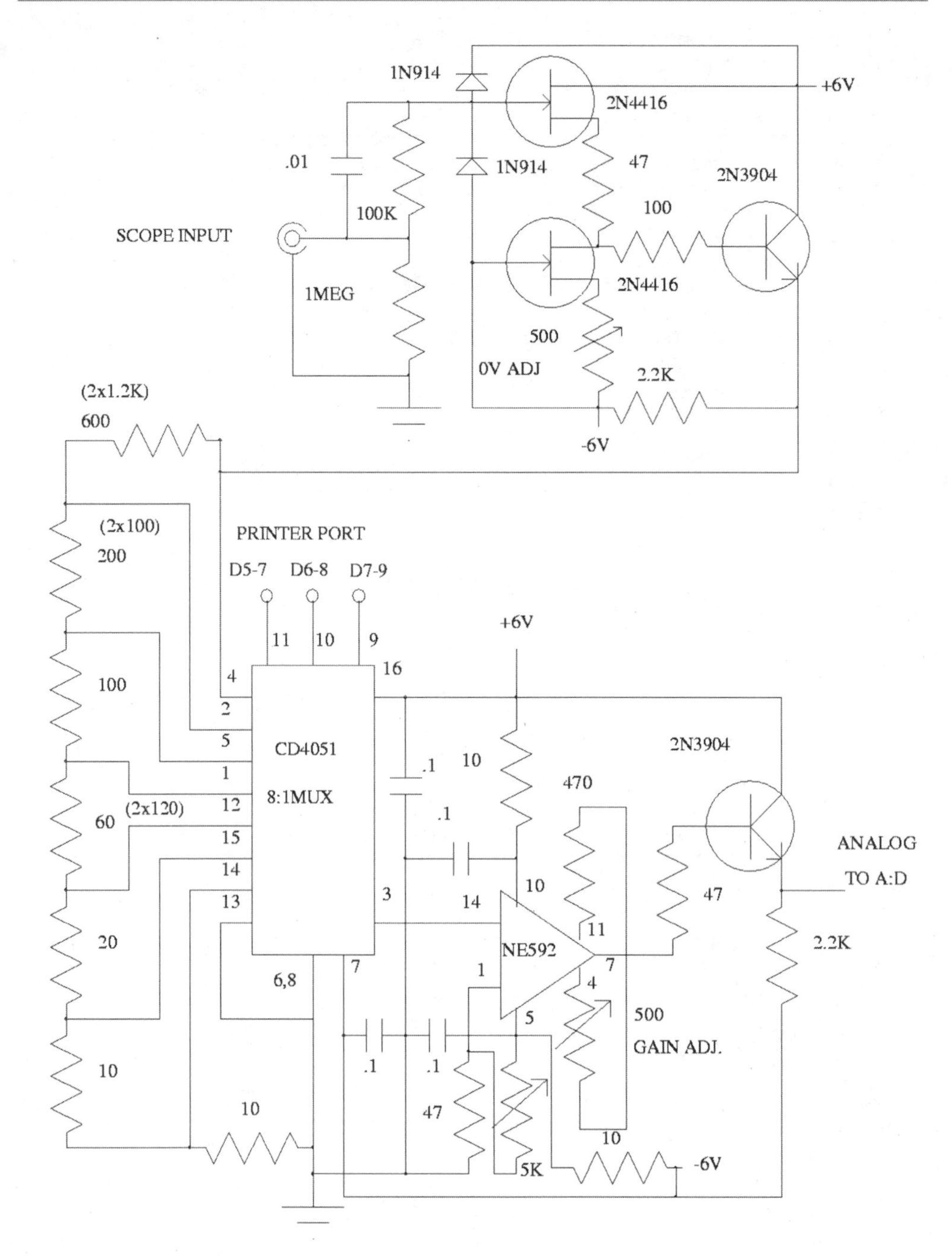

Figure 2-7: Printer port o'scope input section

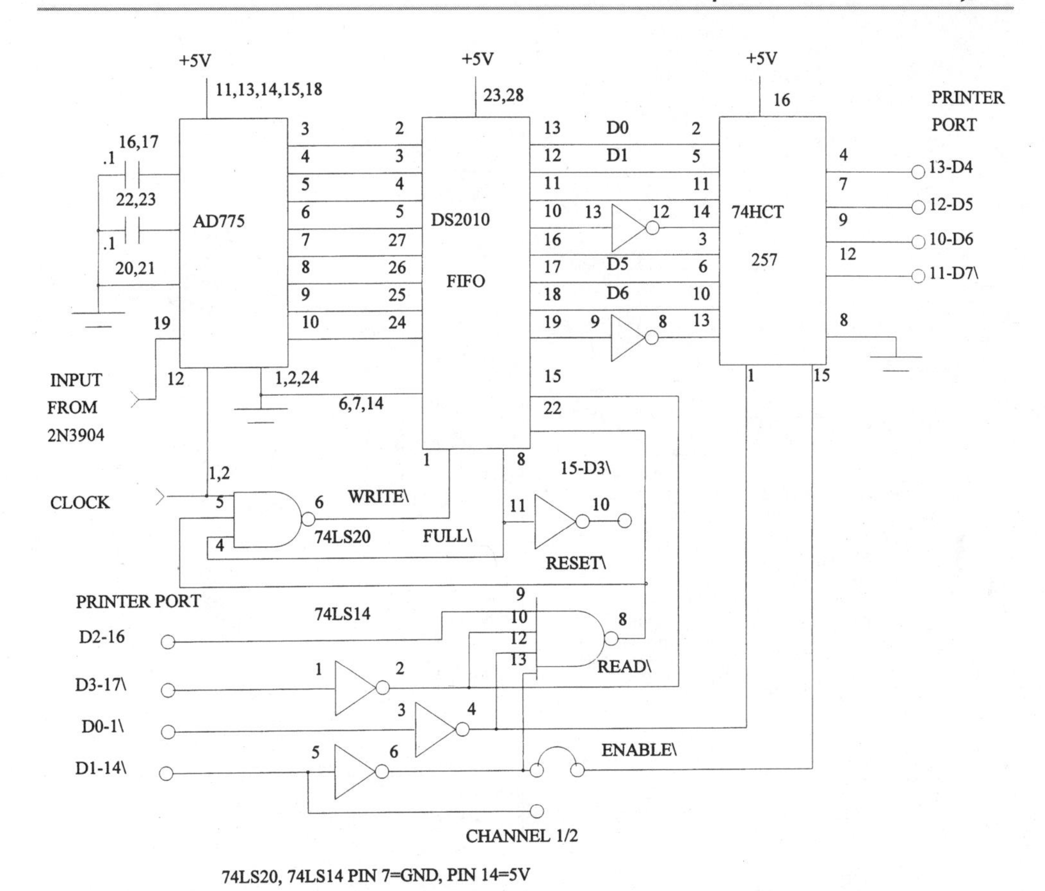

Figure 2-8: Printer port o'scope memory section

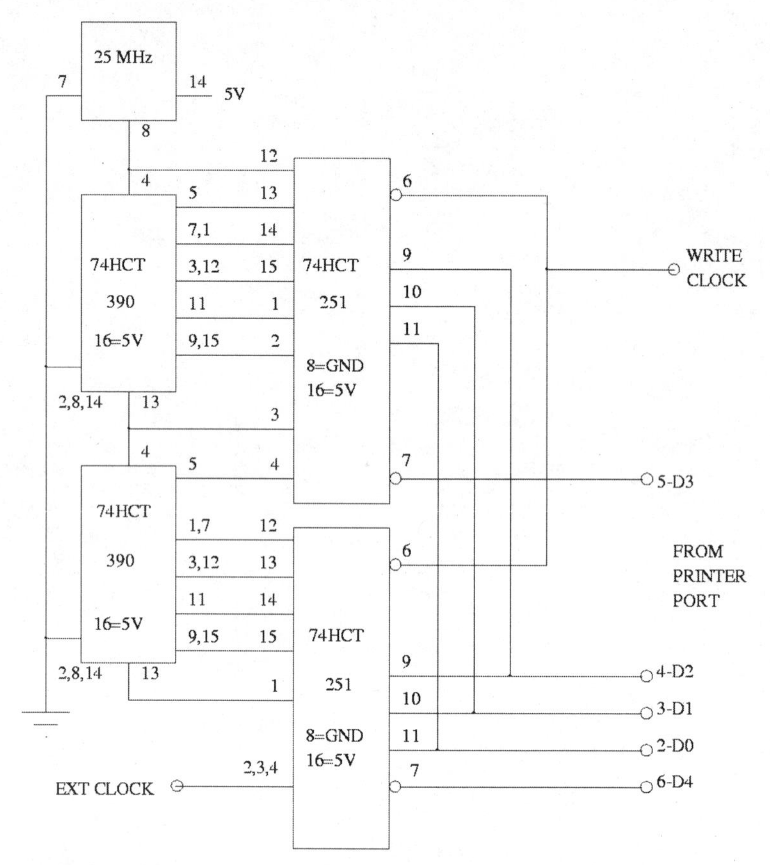

Figure 2-9: Printer port o'scope clock select section

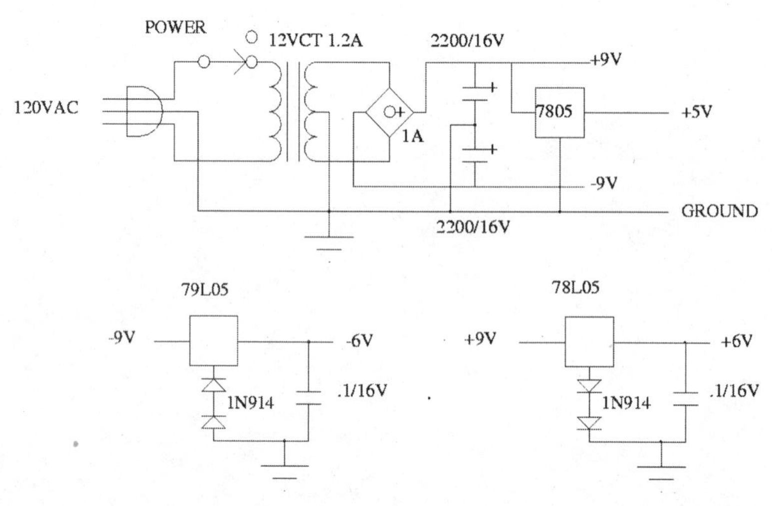

Figure 2-10: Printer port o'scope power supply section

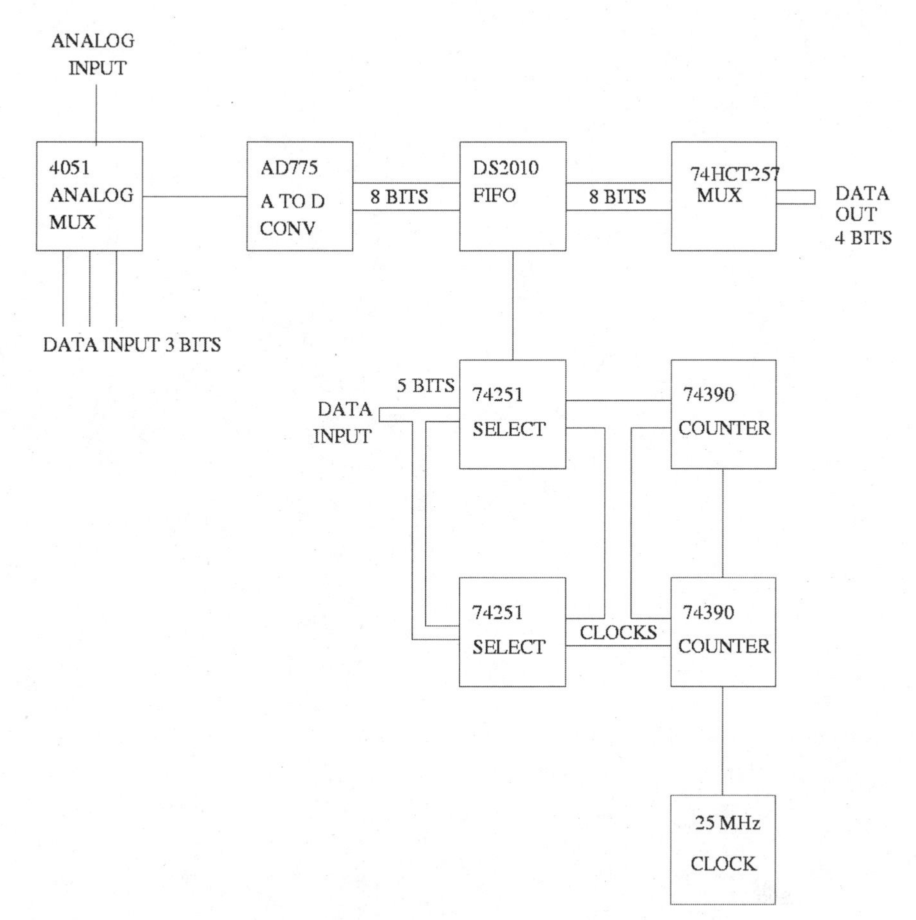

Figure 2-11: Printer port o'scope block diagram

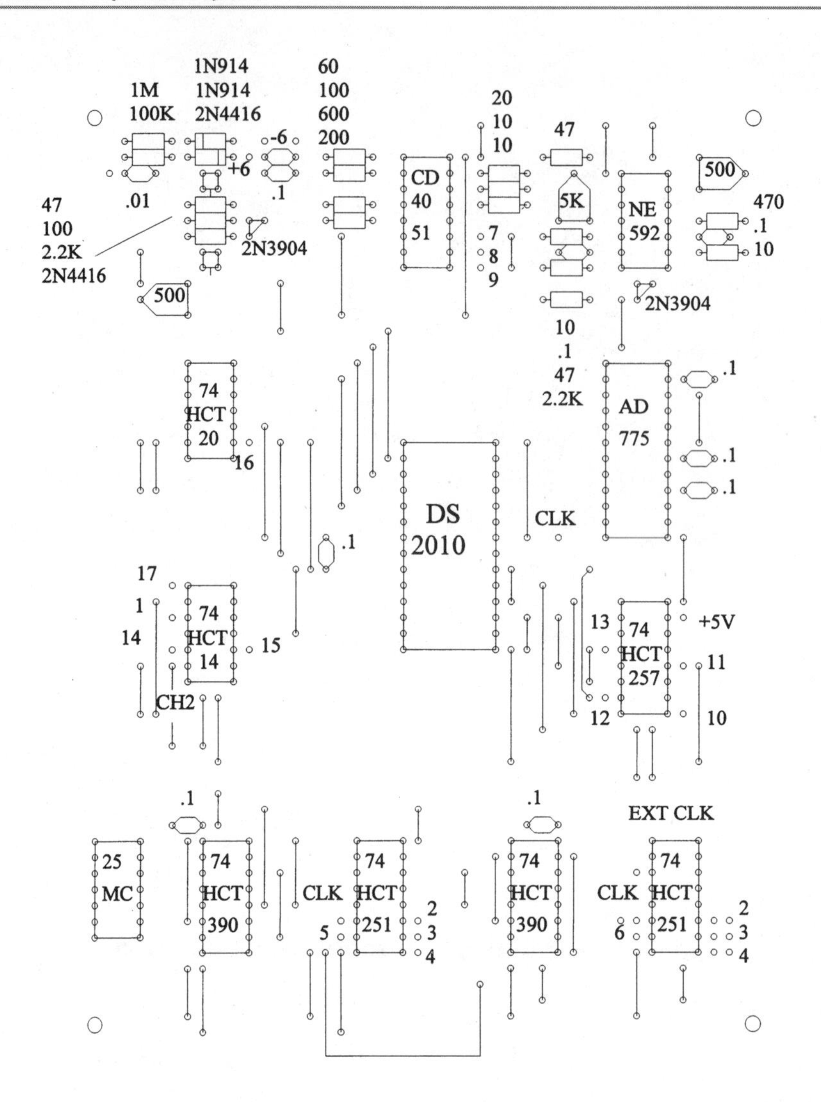

Numbers are pins on the printer port connector

Figure 2-12: Printer port o'scope parts layout

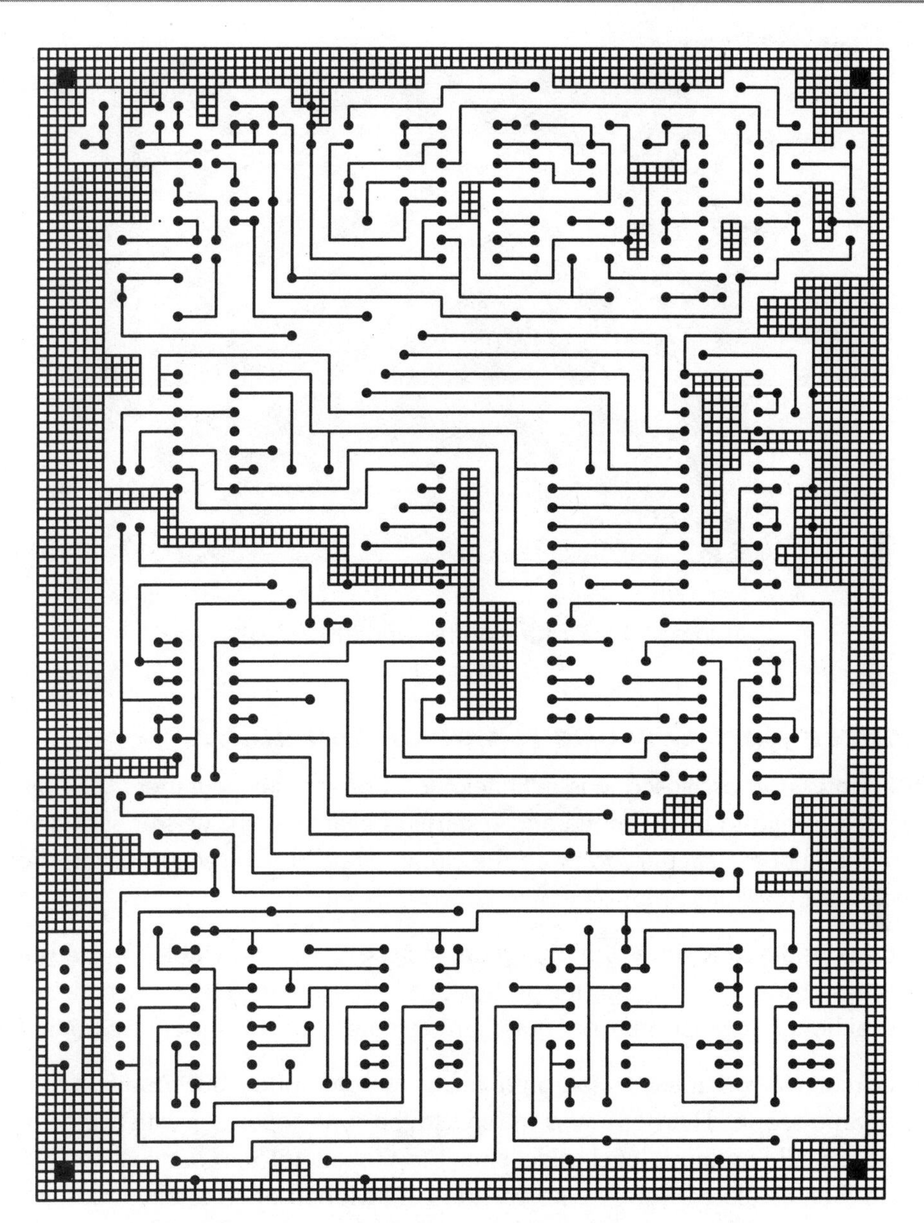

Figure 2-13: Printer port o'scope circuit board

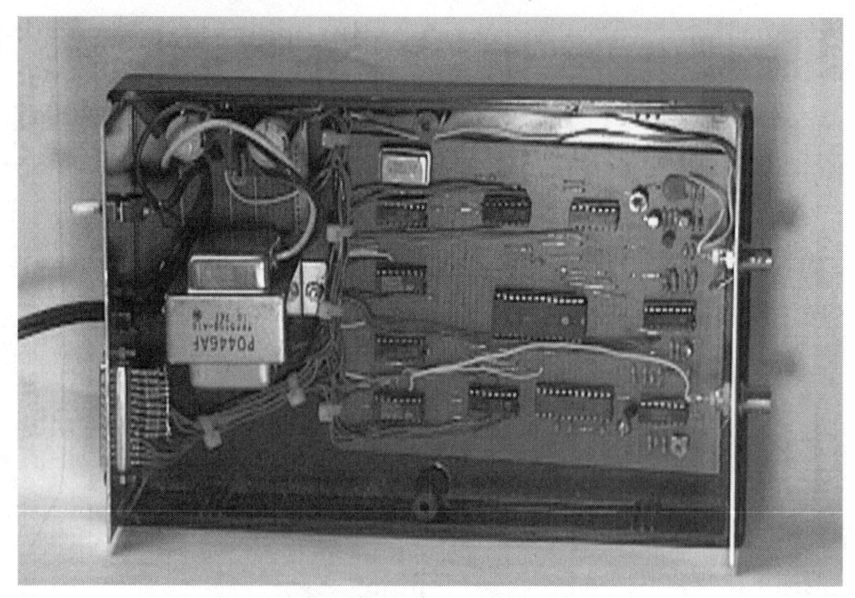

Figure 2-14: Picture of the printer port o'scope

Printer Port O'scope (Dual-Channel) 2-3

This printer port oscilloscope is definitely not a project for beginners. The prototype design had to be scrapped and restarted because there wasn't enough room for all of the circuitry to fit on the circuit board. Each section of this design had several revisions before it would all work together.

This is considered to be a "hobby-grade" device. It is not suitable for medical purposes or accurate enough for professional applications. Design accuracy is 10%, although with laboratory equipment, accuracy of 5% should be obtainable.

This circuit was originally based on a video frame grabber, but modified to make it into an o'scope. The hope was to replace the sync separator with a triggering circuit and add an input level select circuit to make it into an oscilloscope. The design then escalated from earlier versions. It has then been totally redesigned four times. Each time more ICs and circuits were added.

Making a printer port o'scope might appear to be a simple project. Instead it turned out to be almost as complicated as some computers I've designed from scratch. It has been more than 10 years since I've designed anything as complicated. The key components are a high-speed analog-to-digital converter and a high-speed memory device. A new chip from Analog Devices, the

AD775, solves the first problem for $15, rather than the $35 cost of the old 8-bit 15-MHz analog-to-digital converter, the CA3318. The AD775 is available from Newark Electronics, but because of high demand, it might be in short supply.

Bits	Binary	74155 Pin	Mode
0000	0–3	7	Nothing
0100	4	None	Read data channel 1 from 74153s
0101	5	None	Read high 1/2 byte channel 1
0110	6	None	Read data channel 2
0111	7	None	Read high 1/2 byte channel 2
1000	8	9	Write data to channel 1 latch
1001	9	10	Write data to channel 2 latch
1010	10	11	Write data to sweep latch
1011	11	12	Write data to trigger latch
1111	15	4	Reset the o'scope

Table 2-12: Mode selection chart

Position	Frequency	Time
23	25 MHz	2 μs per division
22	10 MHz	5 μs per division
21	5 MHz	10 μs per division
20	2.5 MHz	20 μs per division
19	1 MHz	50 μs per division
18	500 kHz	100 μs per division
17	250 kHz	200 μs per division
16	100 kHz	500 μs per division
15	50 kHz	1 ms per division
14	25 kHz	2 ms per division
13	10 kHz	5 ms per division
12	5 kHz	10 ms per division
11	2.5 kHz	20 ms per division
10	1 kHz	50 ms per division
9	500 Hz	100 ms per division
8	250 Hz	200 ms per division

Table 2-13: Frequency to time conversion

The four problems to deal with when using the AD775 were discussed in the previous section on the first version of the printer port o'scope.

To keep the chip count to a minimum I used a FIFO memory IC instead of SRAM. This is similar to the first version, but this time there are two of them.

The next problem is there are only eight data output bits on a printer port. The first circuit allows the one 8-bit output port to control up to 32 bits. This is the mode selection and latch circuit. It uses the four output control lines to select which one of the four latches will store the output data. It is also used to decode the reset command.

The design is such that it ignores read inputs from the four input control lines used by the input section. This sends data, 4 bits at a time, back to the computer. It also inverts the output control lines to provide true data lines.

Quantity	Number	Pins	Source
2	NE592	14	Newark
2	CD4051	16	Newark
2	AD775	24	Newark (0.4 inches wide)
2	DS2010	28	Newark (or use AM7203A-25PC)
2	74LS153	16	Newark
1	74LS374	20	Newark
1	74LS14	14	Newark
1	74LS20	14	Newark
1	74LS00	14	Newark
1	25-MHz clock	14	Newark
3	73LS390	16	Newark
2	74LS151	16	Newark
3	74LS174	16	Newark
1	74LS157	16	Newark
1	72LS85	16	Newark
1	74LS74	14	Newark
1	74LS10	14	Newark
1	74LS155	16	Newark
1	ULN2003	16	Newark

Note: You can substitute a "HCT" series IC for any of the LS versions.

Table 2-14: 30 ICs used

The clock is very similar to the simplified version, but has three more clock speeds that can be chosen. To divide 25 MHz down to 25 cycles, three of the 74LS390 ICs are necessary to give a total of six divide-by-10s. Then two 74LS151 selectors select the desired clock frequency from eight possibilities each. To choose between the two selectors, data bits 3 and 4 are used. A low data bit enables a 74LS151 selector. The selected clock output is also called the write clock or system clock.

There is also the input gain and attenuation circuit. The first problem is that the frequency select and trigger circuits take up all 8 data bits available on a printer port. To have separate input controls for each channel, additional control data bits are needed for each channel. This is resolved by two data latches, one for each channel located in the mode control circuit.

Quantity	Number	Source
17	16-pin IC sockets	Newark
8	14-pin I.C. sockets	Newark
1	20-pin I.C. socket	Newark
2	24-pin I.C. sockets	Newark
2	28-pin I.C. sockets	Newark
2	BNC connectors	Newark
1	circuit board, 8 × 6 inches	Newark
1	circuit board socket	Newark
1	metal cabinet, 10 × 6 × 2 inches	Newark
1	LM7805 regulator	Newark
1	LM78M05 regulator	Newark
1	LM78L05 regulator	Newark
1	LM79L05 r	Newark
1	25-pin male "D" connector	Newark
4	2N3906	Newark
4	2N4416	Newark
8	1N4148 diodes	Radio Shack
1	12.6 VCT 1.2A transformer	Radio Shack
1	bridge rectifier, 2 amp, 50 V (1 amp would do)	Radio Shack
1	power cord	Radio Shack
1	on/off switch	Radio Shack
Miscellaneous	resistors and capacitors	Radio Shack

Table 2-15: Other materials required

<table>
<tr><td>1</td><td>ground</td><td>analog</td></tr>
<tr><td>2</td><td>+10 V unregulated</td><td>analog</td></tr>
<tr><td>3</td><td>–10 V unregulated</td><td>analog</td></tr>
<tr><td>4</td><td>ground</td><td>analog</td></tr>
<tr><td>5</td><td></td><td></td></tr>
<tr><td>6</td><td></td><td></td></tr>
<tr><td>7</td><td>ground</td><td>digital</td></tr>
<tr><td>8</td><td>pin 1/pin 14</td><td>printer port</td></tr>
<tr><td>9</td><td>pin 2/pin 15</td><td>|</td></tr>
<tr><td>10</td><td>pin 3/pin 16</td><td>|</td></tr>
<tr><td>11</td><td>pin 4/pin 17</td><td>|</td></tr>
<tr><td>12</td><td>pin 5/pin 18</td><td>|</td></tr>
<tr><td>13</td><td>pin 6/pin 19</td><td>|</td></tr>
<tr><td>14</td><td>pin 7/pin 20</td><td>|</td></tr>
<tr><td>15</td><td>pin 8/pin 21</td><td>|</td></tr>
<tr><td>16</td><td>pin 9/pin 22</td><td>|</td></tr>
<tr><td>17</td><td>pin 10/ pin 23</td><td>|</td></tr>
<tr><td>18</td><td>pin 11/ pin 24</td><td>|</td></tr>
<tr><td>19</td><td>pin 12/ pin 25</td><td>|</td></tr>
<tr><td>20</td><td>pin 13</td><td>printer port</td></tr>
<tr><td>21</td><td>+ 5V REGULATED</td><td>digital</td></tr>
<tr><td>22</td><td>ground</td><td>digital</td></tr>
</table>

Table 2-16: Circuit board pinout for 22-pin edge connector

Channel 1	
Volts/division	3 bits gain (D0–2)
Vertical position	Software
Channel 2	
Volts/division	3 bits gain (D0–2)
Vertical position	Software
Sweep	
Time/division	3 bits (D0–2), 2 select bits (D3, D4)
Trigger	
Trigger level	4 bits (D0–3) feeds comparator
Trigger phase	1 bit (D4) selects phase
Trigger channel	1 bit (D5) selects channel

Table 2-17: Controls and how they work

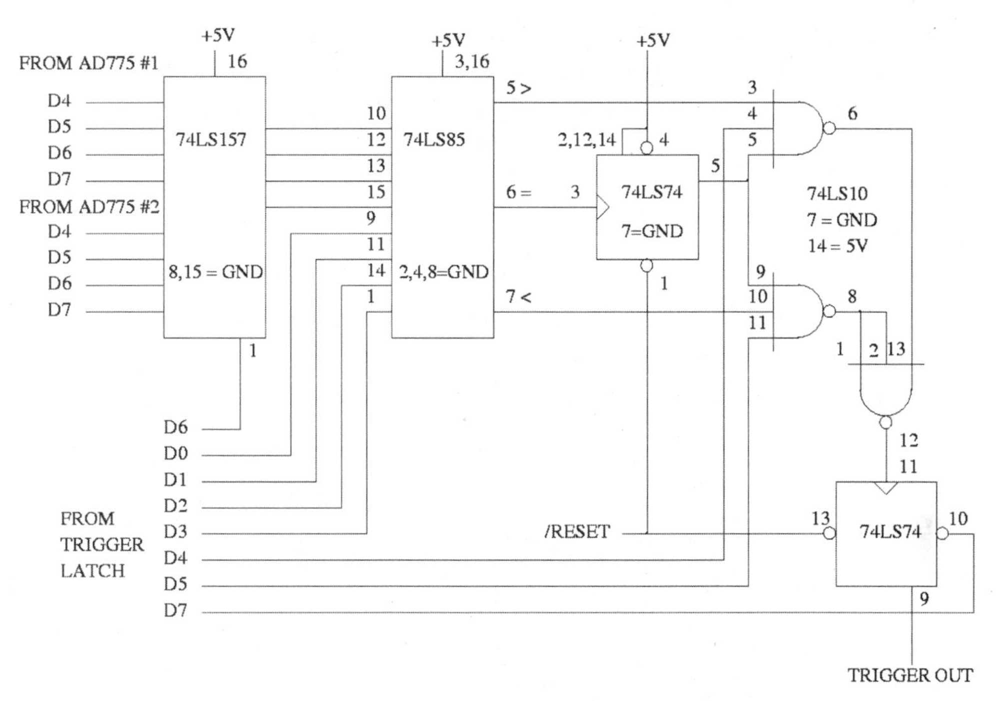

Figure 2-15: Printer port o'scope trigger section

Using data from that channel's latch, a 4051 analog multiplexer selects a tap on a resistor divider to feed the NE592 amplifier. This all works in much the same way as the simplified o'scope, but again with two channels instead of one.

Input attenuation selection and AC or DC coupling selection could be handled with relays, using data bits 3 and 4 for attenuation and bit 5 for AC or DC coupling selection. This optional input attenuation would allow testing of signals greater than the positive and negative 6 VDC power supplies. This power supply limits the entire input section to 12 volts peak-to-peak. By dividing the signal first, inputs of up to 500 volts could be displayed on the oscilloscope. Trimmer capacitors on the input dividers are needed to allow adjusting the frequency response of the attenuator. This is adjusted so that a square wave will be properly displayed.

Next, there is the trigger circuit. A multiplexer allows selection of 4 bits from the channel used as a trigger source. A 74LS85 compares the upper 4 bits of the selected analog-to-digital converter with a trigger value. When the selected value is seen on the output of the analog-to-digital converter, it sets a flipflop.

Then, if the signal under test becomes positive or negative, this will set the "triggered" flipflop. The triggered flipflop then allows sampling to begin. Sampling continues until the FIFO sets its "full" flag. Using 8 bits in the comparison of a trigger would be more precise, but would require even more control bits.

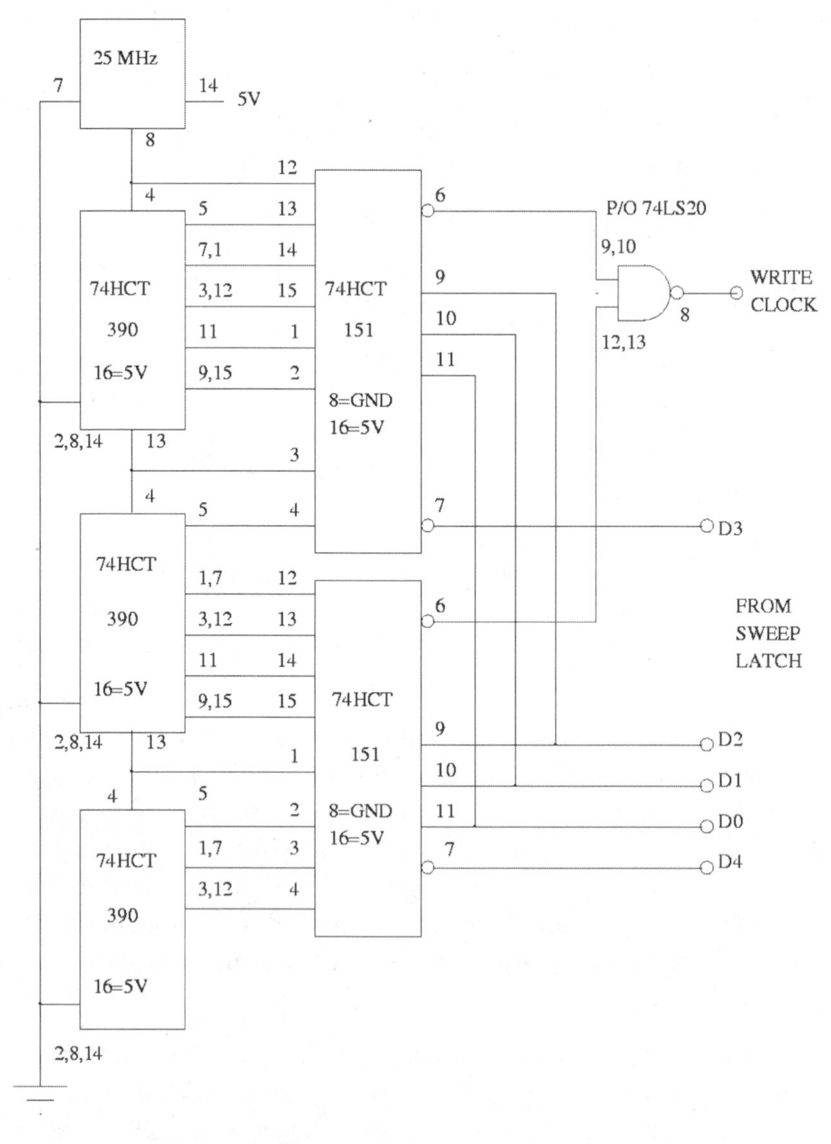

Figure 2-16: Printer port o'scope clock select

The circuit is built on a blank, or a pad per hole 22- or 44-pin edge connector board. Power, ground, and analog ground are created using 20-gauge solid copper wire. After power distribution is connected, check it out with a voltmeter for the proper voltages. Then it is safe to install the ICs.

The NE592s and 4051s are not wire wrapped but rather use point-to-point soldering. They use analog ground buses surrounding their area on the circuit board. The 2N4416s are mounted in this area just above the NE592s.

The analog inputs are connected to the board via 2-pin header connectors. Coax then connects them to the BNC input jacks. The wiring can be done one section at a time in almost any order. The mode latches have to be wired first in order to test the other sections. Each section then is wired and tested with a logic probe or preferably an o'scope to ensure proper operation. The AD775 can be tested with a direct input as long as its voltage remains between ground and 5 volts.

There are two versions of the software, one for CGA and one for VGA resolutions.

```
 ' PRINTER PORT OSCILLOSCOPE
 ' 3/3/95 BY BOB DAVIS
 ' LOW RESOLUTION VERSION FOR 640 X 200 CGA
CLS : SCREEN 2
DRAW "BM 0,0" + "R 639" + "D 190" + "L 639"
DRAW "U 190" + "R 502" + "D 190"
LOCATE 2, 66: PRINT "PRINTER PORT"
LOCATE 3, 66: PRINT "OSCILLOSCOPE "
LOCATE 6, 66: PRINT "CH1 V/DIV:  1"
LOCATE 8, 66: PRINT "CH1 POSIT: 0"
LOCATE 10, 66: PRINT "CH2 V/DIV:  1"
LOCATE 12, 66: PRINT "CH2 POSIT: 0"
LOCATE 14, 66: PRINT "H T/DIV: 5 mS"
LOCATE 16, 66: PRINT "T LEVEL:   8"
LOCATE 18, 66: PRINT "T CH&PH: FREE"
LOCATE 21, 66: PRINT "COPYRIGHT 3/95"
LOCATE 22, 66: PRINT " BY BOB DAVIS"
LOCATE 25, 20: PRINT "SELECT CONTROLS WITH ARROW KEYS OR Q=QUIT";
T = 13: S1 = 2: S2 = 2: P1 = 0: P2 = 0: C = 6: R = 8: TM = 7
START: OUT &H37A, 11: OUT &H378, T           'TIME BASE
OUT &H37A, 10: OUT &H378, (R + (TM * 16))    'TRIGGER
OUT &H37A, 9: OUT &H378, S2                   'SENSITIVITY CH2
OUT &H37A, 8: OUT &H378, S1                   'SENSITIVITY CH1
OUT &H37A, 15                                  'RESET - START
FOR A = 1 TO 1000: NEXT A                     'CAPTURE DELAY
FOR H = 1 TO 500                               'HORIZONTAL RANGE
PSET (H, 2), 0: DRAW "C0" + "M=" + VARPTR$(H) + ",189"
IF H > 2 THEN PSET (H - 1, V1), 0 ELSE PSET (H, 95), 0
OUT &H37A, 4: V1 = (INP(&H379) \ 16)      'GET LOW 1/2 BYTE
```

```
OUT &H37A, 5: V1 = V1 + (INP(&H379) AND &HF0) - P1 - 30
IF V1 > 188 THEN V1 = 188 ELSE IF V1 < 2 THEN V1 = 2
IF H > 2 THEN DRAW "C1" + "M=" + VARPTR$(H) + ",=" + VARPTR$(V1)
IF H > 2 THEN PSET (H - 1, V2), 0 ELSE PSET (H, 95), 0
OUT &H37A, 6: V2 = (INP(&H379) \ 16)      'GET LOW 1/2 BYTE
OUT &H37A, 7: V2 = V2 + (INP(&H379) AND &HF0) - P2 - 30
IF V2 > 188 THEN V2 = 188 ELSE IF V2 < 2 THEN V2 = 2
IF H > 2 THEN DRAW "C1" + "M=" + VARPTR$(H) + ",=" + VARPTR$(V2)
OUT &H37A, 0                                'NEXT ADDRESS
NEXT H
key$ = INKEY$: IF key$ = "q" OR key$ = "Q" THEN END
LOCATE C, 64: PRINT "  "
IF key$ = "2" THEN C = C + 2: IF C = 20 THEN C = 6
IF key$ = "8" THEN C = C - 2: IF C = 4 THEN C = 18
LOCATE C, 64: PRINT "->"
IF C = 6 THEN
IF key$ = "6" THEN : S1 = S1 + 1: IF S1 > 7 THEN S1 = 0
IF key$ = "4" THEN : S1 = S1 - 1: IF S1 < 0 THEN S1 = 7
IF S1 = 0 THEN LOCATE 6, 76: PRINT "GND"
IF S1 = 1 THEN LOCATE 6, 76: PRINT "  2"
IF S1 = 2 THEN LOCATE 6, 76: PRINT "  1"
IF S1 = 3 THEN LOCATE 6, 76: PRINT " .5"
IF S1 = 4 THEN LOCATE 6, 76: PRINT " .2"
IF S1 = 5 THEN LOCATE 6, 76: PRINT " .1"
IF S1 = 6 THEN LOCATE 6, 76: PRINT ".05"
IF S1 = 7 THEN LOCATE 6, 76: PRINT ".02"
END IF
IF C = 8 THEN
IF key$ = "6" THEN : P1 = P1 + 2: IF P1 > 70 THEN P1 = -70
IF key$ = "4" THEN : P1 = P1 - 2: IF P1 < -70 THEN P1 = 70
LOCATE 8, 76: PRINT P1
END IF
IF C = 10 THEN
IF key$ = "6" THEN : S2 = S2 + 1: IF S2 > 7 THEN S2 = 0
IF key$ = "4" THEN : S2 = S2 - 1: IF S2 < 0 THEN S2 = 7
IF S2 = 0 THEN LOCATE 10, 76: PRINT "GND"
IF S2 = 1 THEN LOCATE 10, 76: PRINT "  2"
IF S2 = 2 THEN LOCATE 10, 76: PRINT "  1"
IF S2 = 3 THEN LOCATE 10, 76: PRINT " .5"
IF S2 = 4 THEN LOCATE 10, 76: PRINT " .2"
IF S2 = 5 THEN LOCATE 10, 76: PRINT " .1"
IF S2 = 6 THEN LOCATE 10, 76: PRINT ".05"
IF S2 = 7 THEN LOCATE 10, 76: PRINT ".02"
END IF
IF C = 12 THEN
IF key$ = "6" THEN : P2 = P2 + 2: IF P2 > 70 THEN P2 = -70
IF key$ = "4" THEN : P2 = P2 - 2: IF P2 < -70 THEN P2 = 70
LOCATE 12, 76: PRINT P2
END IF
IF C = 14 THEN
IF key$ = "6" THEN : T = T + 1: IF T > 23 THEN T = 8
IF key$ = "4" THEN : T = T - 1: IF T < 8 THEN T = 23
IF T = 8 THEN LOCATE 14, 74: PRINT " .2 S"
```

```
IF T = 9 THEN LOCATE 14, 74: PRINT " .1 S"
IF T = 10 THEN LOCATE 14, 74: PRINT "50 mS"
IF T = 11 THEN LOCATE 14, 74: PRINT "20 mS"
IF T = 12 THEN LOCATE 14, 74: PRINT "10 mS"
IF T = 13 THEN LOCATE 14, 74: PRINT " 5 mS"
IF T = 14 THEN LOCATE 14, 74: PRINT " 2 mS"
IF T = 15 THEN LOCATE 14, 74: PRINT " 1 mS"
IF T = 16 THEN LOCATE 14, 74: PRINT ".5 mS"
IF T = 17 THEN LOCATE 14, 74: PRINT ".2 mS"
IF T = 18 THEN LOCATE 14, 74: PRINT ".1 mS"
IF T = 19 THEN LOCATE 14, 74: PRINT "50 uS"
IF T = 20 THEN LOCATE 14, 74: PRINT "20 uS"
IF T = 21 THEN LOCATE 14, 74: PRINT "10 uS"
IF T = 22 THEN LOCATE 14, 74: PRINT " 5 uS"
IF T = 23 THEN LOCATE 14, 74: PRINT " 2 uS"
END IF
IF C = 16 THEN
IF key$ = "6" THEN : R = R + 1: IF R > 15 THEN R = 0
IF key$ = "4" THEN : R = R - 1: IF R < 0 THEN R = 15
LOCATE 16, 76: PRINT R
END IF
IF C = 18 THEN
IF key$ = "6" THEN : TM = TM + 1: IF TM > 15 THEN TM = 7
IF key$ = "4" THEN : TM = TM - 1: IF TM < 7 THEN TM = 15
IF TM < 8 THEN LOCATE 18, 74: PRINT " FREE"
IF TM = 8 THEN LOCATE 18, 74: PRINT " NONE"
IF TM = 9 THEN LOCATE 18, 74: PRINT "CH 1-"
IF TM = 10 THEN LOCATE 18, 74: PRINT "CH 1+"
IF TM = 11 THEN LOCATE 18, 74: PRINT "CH 1="
IF TM = 12 THEN LOCATE 18, 74: PRINT " NONE"
IF TM = 13 THEN LOCATE 18, 74: PRINT "CH 2-"
IF TM = 14 THEN LOCATE 18, 74: PRINT "CH 2+"
IF TM = 15 THEN LOCATE 18, 74: PRINT "CH 2="
END IF
GOTO START
```

```
' PRINTER PORT OSCILLOSCOPE 3/3/95 BY BOB DAVIS
' HIGH RESOLUTION VERSION FOR 640 X 350 EGA/VGA
' SET GAIN NEAR MAX, CHANGED FROM 200 V STEPS TO 350.
INPUT "USE PRINTER PORT NUMBER? ", LPT
' DOUT=DATA OUT, COUT=CONTROL OUT, CIN=CONTROL IN
IF LPT > 0 AND LPT < 4 THEN
IF LPT = 1 THEN DOUT = &H3BC: COUT = &H3BE: CIN = &H3BD
IF LPT = 2 THEN DOUT = &H378: COUT = &H37A: CIN = &H379
IF LPT = 3 THEN DOUT = &H278: COUT = &H27A: CIN = &H279
ELSE
DOUT = &H378: COUT = &H37A: CIN = &H379
END IF
CLS : SCREEN 9
DRAW "BM 0,0" + "R 639" + "D 330" + "L 639"
DRAW "U 330" + "R 502" + "D 330"
COLOR 4
PSET (0, 30): PSET (0, 60): PSET (0, 90): PSET (0, 120): PSET (0, 150)
```

```
PSET (0, 180): PSET (0, 210): PSET (0, 240): PSET (0, 270): PSET (0, 300)
PSET (502, 30): PSET (502, 60): PSET (502, 90): PSET (502, 120)
PSET (502, 150): PSET (502, 180): PSET (502, 210): PSET (502, 240)
PSET (502, 270): PSET (502, 300)
PSET (50, 0): PSET (100, 0): PSET (150, 0): PSET (200, 0): PSET (250, 0)
PSET (300, 0): PSET (350, 0): PSET (400, 0): PSET (450, 0): PSET (50, 330)
PSET (100, 330): PSET (150, 330): PSET (200, 330): PSET (250, 330)
PSET (300, 330): PSET (350, 330): PSET (400, 330): PSET (450, 330)
COLOR 2
LOCATE 2, 66: PRINT "PRINTER PORT"
LOCATE 3, 66: PRINT "OSCILLOSCOPE "
LOCATE 6, 66: PRINT "CH1 V/DIV:  1"
LOCATE 8, 66: PRINT "CH1 POSIT: 0"
LOCATE 10, 66: PRINT "CH2 V/DIV:  1"
LOCATE 12, 66: PRINT "CH2 POSIT: 0"
LOCATE 14, 66: PRINT "H T/DIV: 5 mS"
LOCATE 16, 66: PRINT "T LEVEL:   8"
LOCATE 18, 66: PRINT "T CH&PH: FREE"
LOCATE 21, 66: PRINT "COPYRIGHT 3/95"
LOCATE 22, 66: PRINT " BY BOB DAVIS"
LOCATE 25, 20: PRINT "SELECT CONTROLS WITH ARROW KEYS OR Q=QUIT";
T = 13: S1 = 2: S2 = 2: P1 = 0: P2 = 0: C = 6: R = 8: TM = 7
START: OUT COUT, 11: OUT DOUT, T                    'TIME BASE
OUT COUT, 10: OUT DOUT, (R + (TM * 16))            'TRIGGER
OUT COUT, 9: OUT DOUT, S2                           'SENSITIVITY CH2
OUT COUT, 8: OUT DOUT, S1                           'SENSITIVITY CH1
OUT COUT, 15                                       'RESET - START
FOR A = 1 TO 1000: NEXT A                        'CAPTURE DELAY
FOR H = 1 TO 500                                  'HORIZONTAL RANGE
PSET (H, 2), 0: DRAW "C0" + "M=" + VARPTR$(H) + ",329"
IF H > 2 THEN PSET (H - 1, V1), 0 ELSE PSET (H, 95), 0
OUT COUT, 4: V1 = (INP(CIN) \ 16)               'GET LOW 1/2 BYTE
OUT COUT, 5: V1 = V1 + (INP(CIN) AND &HF0) - P1
IF V1 < 2 THEN V1 = 2: IF V1 > 328 THEN V1 = 328
IF H > 2 THEN DRAW "C3" + "M=" + VARPTR$(H) + ",=" + VARPTR$(V1)
IF H > 2 THEN PSET (H - 1, V2), 0 ELSE PSET (H, 95), 0
OUT COUT, 6: V2 = (INP(CIN) \ 16)               'GET LOW 1/2 BYTE
OUT COUT, 7: V2 = V2 + (INP(CIN) AND &HF0) - P2
IF V2 < 2 THEN V2 = 2: IF V2 > 328 THEN V2 = 328
IF H > 2 THEN DRAW "C2" + "M=" + VARPTR$(H) + ",=" + VARPTR$(V2)
OUT COUT, 0: OUT COUT, 8                         'NEXT ADDRESS
NEXT H
key$ = INKEY$: IF key$ = "q" OR key$ = "Q" THEN END
LOCATE C, 64: PRINT "  "
IF key$ = "2" THEN C = C + 2: IF C = 20 THEN C = 6
IF key$ = "8" THEN C = C - 2: IF C = 4 THEN C = 18
COLOR 4: LOCATE C, 64: PRINT "->": COLOR 2
IF C = 6 THEN
IF key$ = "6" THEN : S1 = S1 + 1: IF S1 > 7 THEN S1 = 0
IF key$ = "4" THEN : S1 = S1 - 1: IF S1 < 0 THEN S1 = 7
IF S1 = 0 THEN LOCATE 6, 76: PRINT "GND"
IF S1 = 1 THEN LOCATE 6, 76: PRINT "  2"
IF S1 = 2 THEN LOCATE 6, 76: PRINT "  1"
```

```
IF S1 = 3 THEN LOCATE 6, 76: PRINT " .5"
IF S1 = 4 THEN LOCATE 6, 76: PRINT " .2"
IF S1 = 5 THEN LOCATE 6, 76: PRINT " .1"
IF S1 = 6 THEN LOCATE 6, 76: PRINT ".05"
IF S1 = 7 THEN LOCATE 6, 76: PRINT ".02"
END IF
IF C = 8 THEN
IF key$ = "6" THEN : P1 = P1 + 2: IF P1 > 70 THEN P1 = -70
IF key$ = "4" THEN : P1 = P1 - 2: IF P1 < -70 THEN P1 = 70
LOCATE 8, 76: PRINT P1
END IF
IF C = 10 THEN
IF key$ = "6" THEN : S2 = S2 + 1: IF S2 > 7 THEN S2 = 0
IF key$ = "4" THEN : S2 = S2 - 1: IF S2 < 0 THEN S2 = 7
IF S2 = 0 THEN LOCATE 10, 76: PRINT "GND"
IF S2 = 1 THEN LOCATE 10, 76: PRINT "  2"
IF S2 = 2 THEN LOCATE 10, 76: PRINT "  1"
IF S2 = 3 THEN LOCATE 10, 76: PRINT " .5"
IF S2 = 4 THEN LOCATE 10, 76: PRINT " .2"
IF S2 = 5 THEN LOCATE 10, 76: PRINT " .1"
IF S2 = 6 THEN LOCATE 10, 76: PRINT ".05"
IF S2 = 7 THEN LOCATE 10, 76: PRINT ".02"
END IF
IF C = 12 THEN
IF key$ = "6" THEN : P2 = P2 + 2: IF P2 > 70 THEN P2 = -70
IF key$ = "4" THEN : P2 = P2 - 2: IF P2 < -70 THEN P2 = 70
LOCATE 12, 76: PRINT P2
END IF
IF C = 14 THEN
IF key$ = "6" THEN : T = T + 1: IF T > 23 THEN T = 8
IF key$ = "4" THEN : T = T - 1: IF T < 8 THEN T = 23
IF T = 8 THEN LOCATE 14, 74: PRINT " .2 S"
IF T = 9 THEN LOCATE 14, 74: PRINT " .1 S"
IF T = 10 THEN LOCATE 14, 74: PRINT "50 mS"
IF T = 11 THEN LOCATE 14, 74: PRINT "20 mS"
IF T = 12 THEN LOCATE 14, 74: PRINT "10 mS"
IF T = 13 THEN LOCATE 14, 74: PRINT " 5 mS"
IF T = 14 THEN LOCATE 14, 74: PRINT " 2 mS"
IF T = 15 THEN LOCATE 14, 74: PRINT " 1 mS"
IF T = 16 THEN LOCATE 14, 74: PRINT ".5 mS"
IF T = 17 THEN LOCATE 14, 74: PRINT ".2 mS"
IF T = 18 THEN LOCATE 14, 74: PRINT ".1 mS"
IF T = 19 THEN LOCATE 14, 74: PRINT "50 uS"
IF T = 20 THEN LOCATE 14, 74: PRINT "20 uS"
IF T = 21 THEN LOCATE 14, 74: PRINT "10 uS"
IF T = 22 THEN LOCATE 14, 74: PRINT " 5 uS"
IF T = 23 THEN LOCATE 14, 74: PRINT " 2 uS"
END IF
IF C = 16 THEN
IF key$ = "6" THEN : R = R + 1: IF R > 15 THEN R = 0
IF key$ = "4" THEN : R = R - 1: IF R < 0 THEN R = 15
LOCATE 16, 76: PRINT R
END IF
```

```
IF C = 18 THEN
IF key$ = "6" THEN : TM = TM + 1: IF TM > 15 THEN TM = 7
IF key$ = "4" THEN : TM = TM - 1: IF TM < 7 THEN TM = 15
IF TM < 8 THEN LOCATE 18, 74: PRINT " FREE"
IF TM = 8 THEN LOCATE 18, 74: PRINT " NONE"
IF TM = 9 THEN LOCATE 18, 74: PRINT "CH 1-"
IF TM = 10 THEN LOCATE 18, 74: PRINT "CH 1+"
IF TM = 11 THEN LOCATE 18, 74: PRINT "CH 1="
IF TM = 12 THEN LOCATE 18, 74: PRINT " NONE"
IF TM = 13 THEN LOCATE 18, 74: PRINT "CH 2-"
IF TM = 14 THEN LOCATE 18, 74: PRINT "CH 2+"
IF TM = 15 THEN LOCATE 18, 74: PRINT "CH 2="
END IF
GOTO START
```

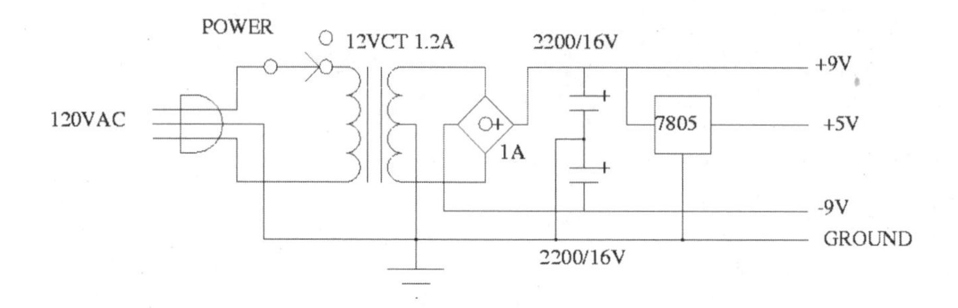

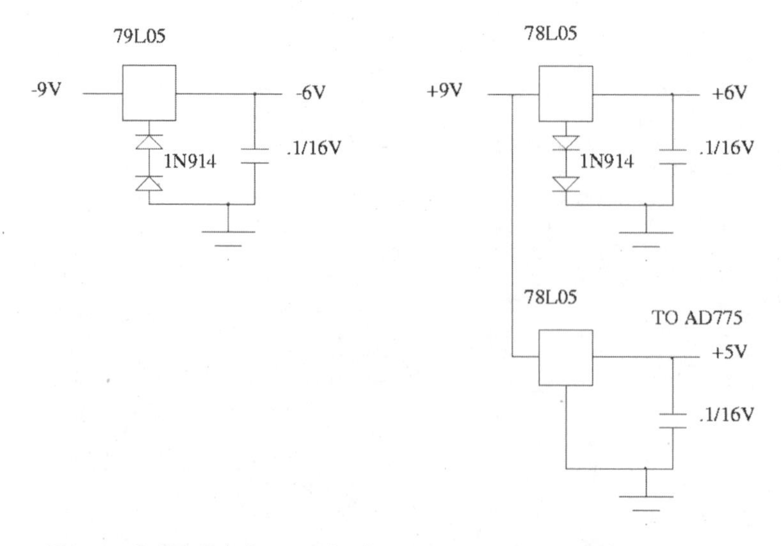

Figure 2-17: Printer port o'scope power supply

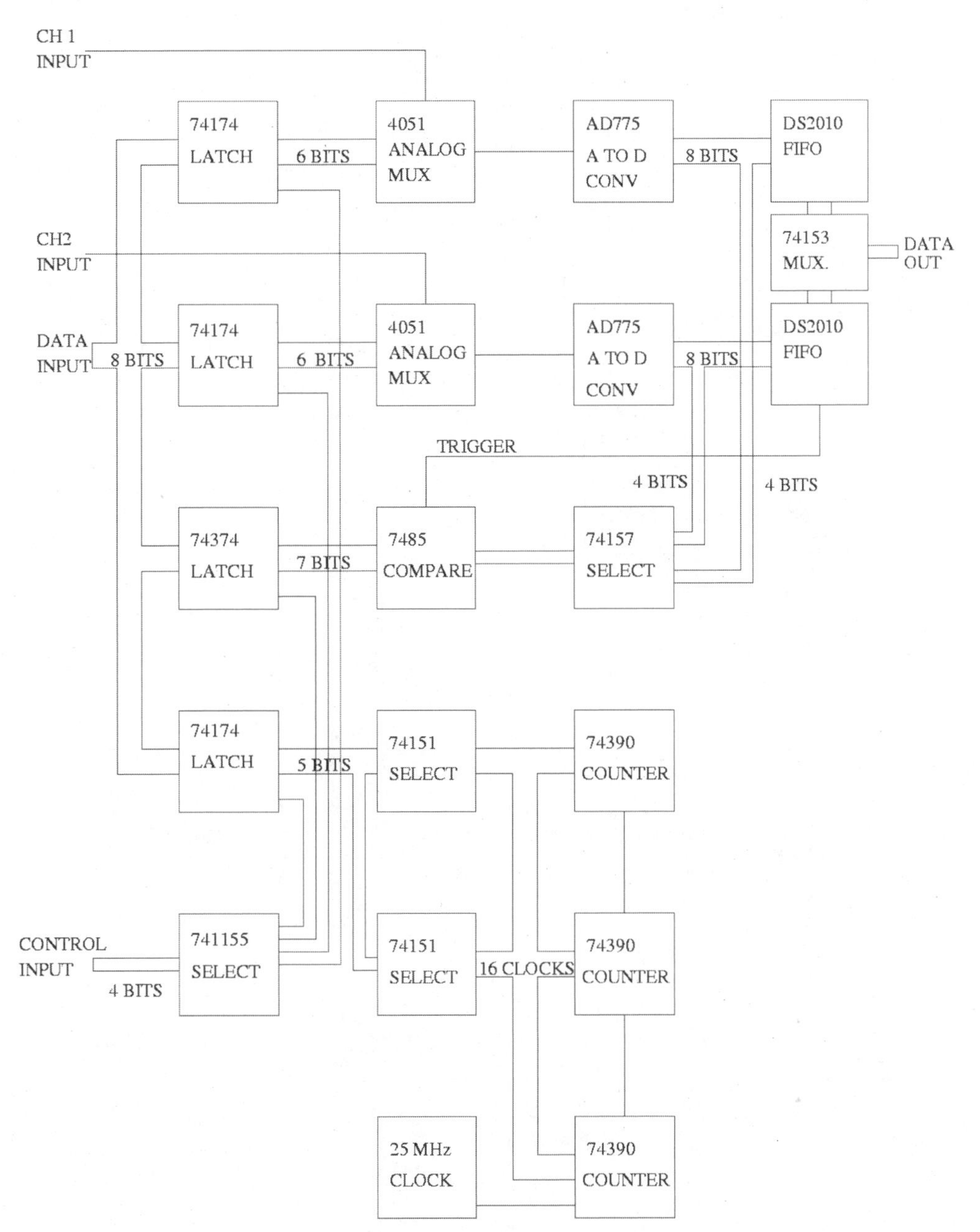

Figure 2-18: Printer port o'scope block diagram

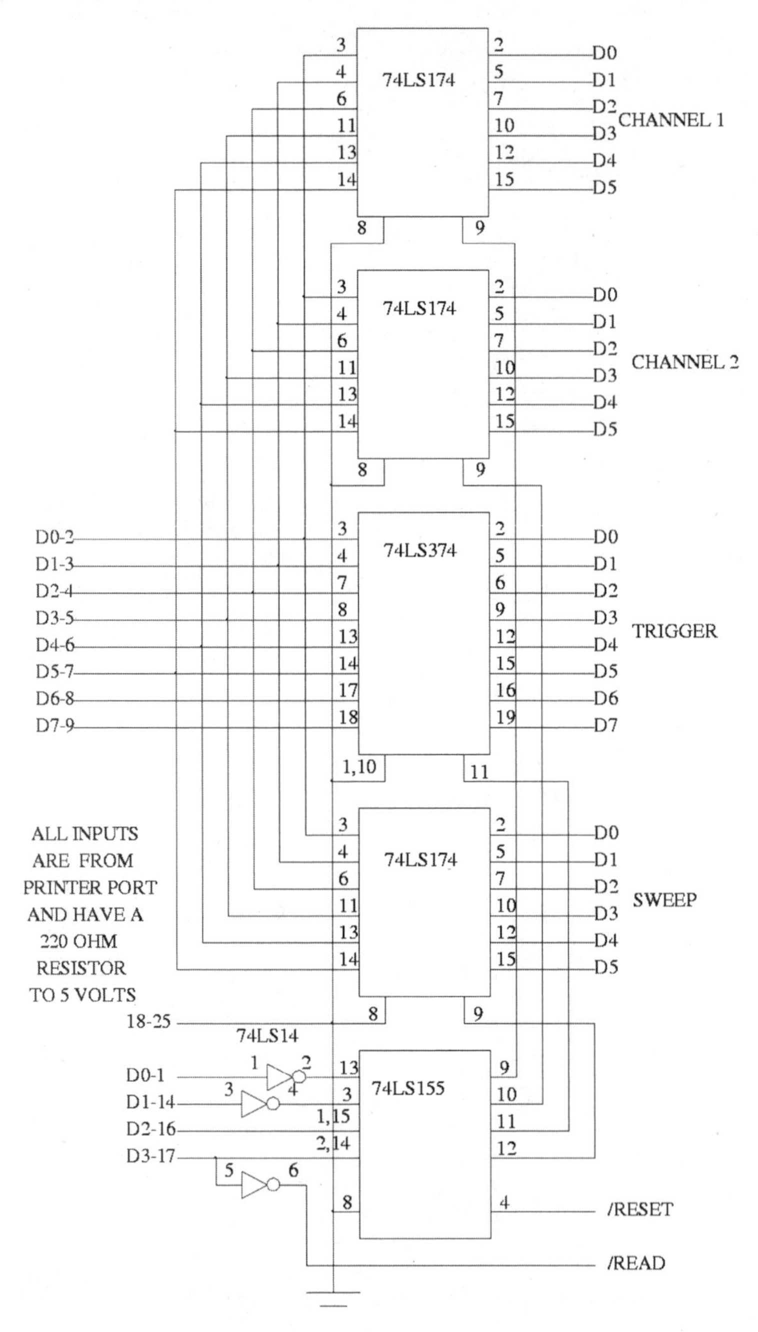

Figure 2-19: Printer port o'scope latches

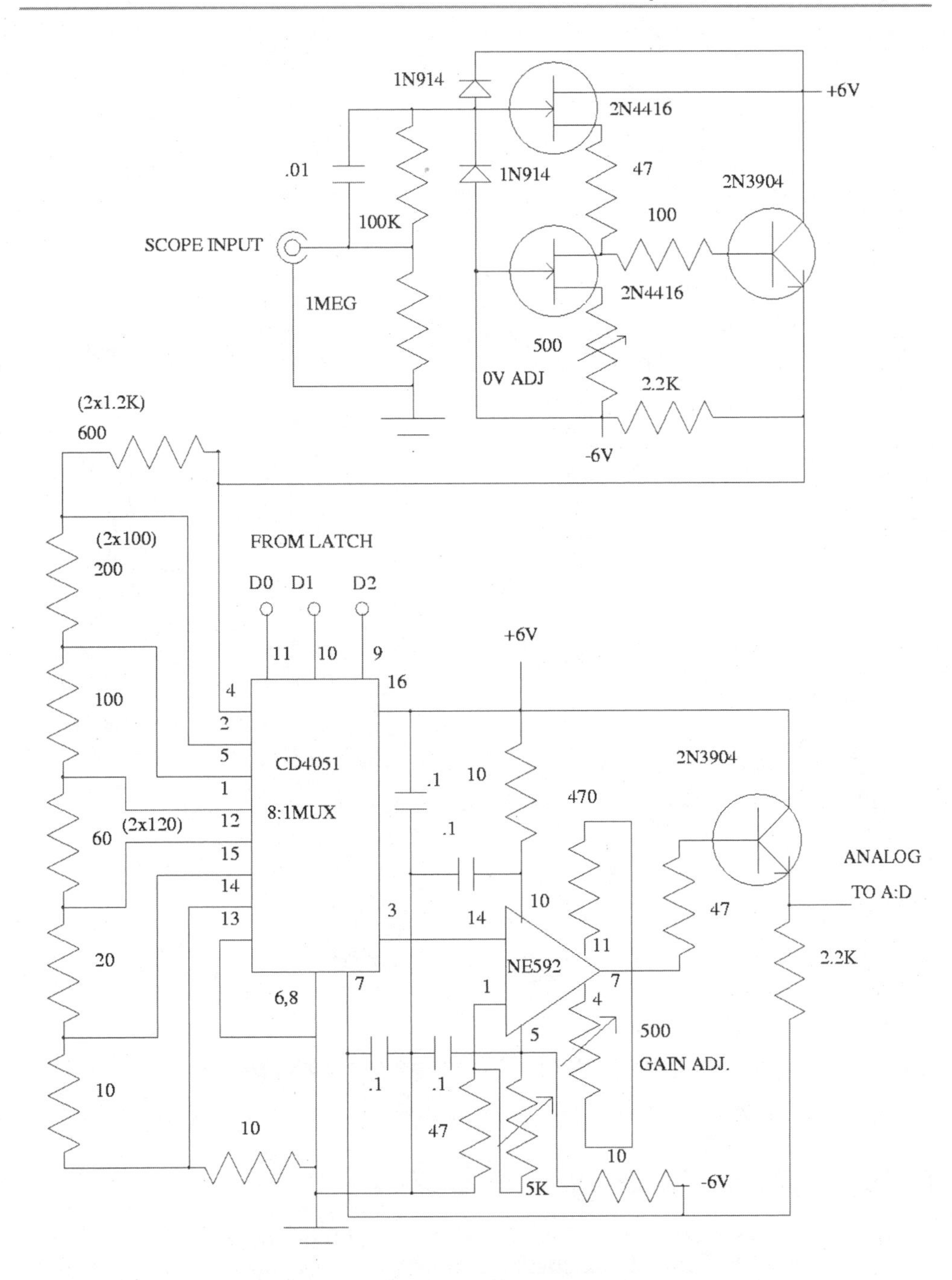

Figure 2-20: Printer port o'scope input section

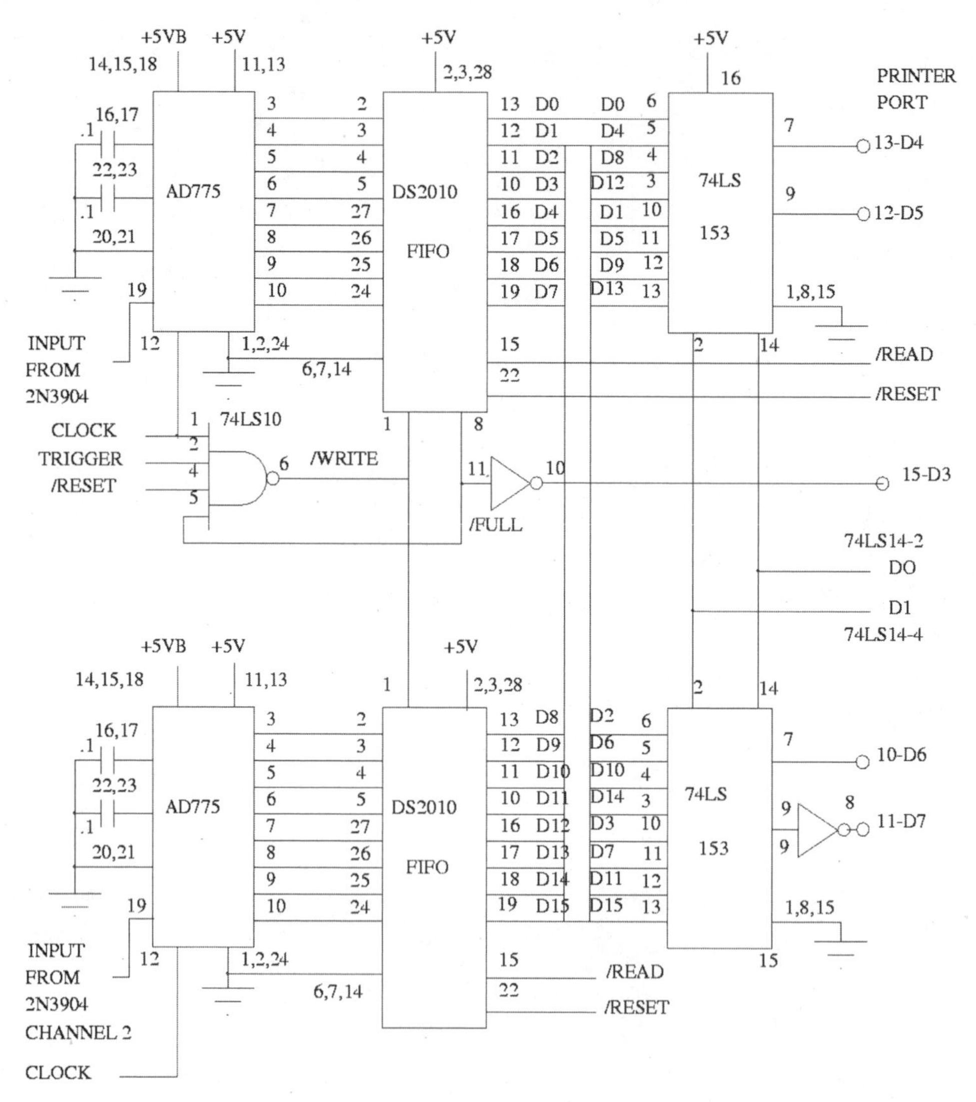

Figure 2-21: Printer port o'scope memory section

Printer Port Audio Analyzer 2-4

This device grew out of another project that was something of a failure. I wanted to make a sound mixer that was totally computerized. The best I could do was an eight-channel, eight-control-per-channel version. The right ICs were just not available for that purpose. The closest available is the MAX528, an eight-channel digital-to-analog converter, and the TDA1524, which was designed for car radios. The TDA1524 supplies analog-controlled bass, treble, balance, and volume on one IC. These two ICs work well for a DJ-type mixer, but are not suitable for professional sound mixers.

One thing I did develop, however, is the ability to monitor volume levels for all channels of a mixer. This can be displayed on a computer monitor. Taking this feature and adding it to a normal mixer gives the ability to view activity for all channels of the mixer. In theory, the design could be expanded to add the ability to give a 15-frequency spectrum analyzer. You could even make a 32-channel version that can monitor 32 channels and display a 30-frequency spectrum analyzer. The two could also be crossed, providing 16 VU meters and a 15-channel spectrum analyzer on the same screen.

Many inexpensive mixers do not have a channel activity light. The most expensive mixers made do not have a 32-step VU meter on every channel. But with the printer port audio analyzer, you can add one to almost any mixer.

For this project we will consider a scaled-down version. We will discuss a 16- or 32-channel VU meter that can be added to any mixer with a "tap," "insert," or "line out" on each channel. Each of the 16 or 32 channels have 32 steps displayed on the screen. This is the equivalent to 512, or even 1,024 LEDs.

To keep the price at rock bottom, a discarded 286-based computer can be used. The software had to be optimized and compiled to run on a slow 286 computer. CGA graphics and a 12-inch color monitor work fine, but a black-and-white monitor would work as well. Salvaging everything from the spare parts pile, except some LF353s, 1/4-inch plugs, and RCA jacks, kept the project under $20. If everything was purchased new, the price should still be under $100, except for the computer.

The basic computer-controlled audio analyzer consists of 16 full-wave rectifiers and a 16-port analog-to-digital converter. The converter is controlled by the computer's printer port. The computer then checks each channel in sequence and displays the acquired levels on the screen.

The full-wave rectifier's response time is controlled by the resistor feeding the 1 μF capacitor. For a 486 computer, 10K works fine, but for slower 286s you will need to use a 47K or 100K resistor. This will limit the display to the average, not the peak level. To save space and use parts on hand, I used some 10K resistor arrays on the incoming lines. The LF353 can be substituted by a TL082 or perhaps even one of the old LM1558s. The diodes aren't critical, either.

The CD4051 is an 8-channel-to-one-channel analog multiplexer. Two are used for 16 channels, and 4 would be needed for 32 channels. The MAX154 is an easy-to-use four-to-one multiplexer, 8-bit analog-to-digital converter, and voltage reference in one small IC. By using four CD4051s, the design is easily expanded to 32 channels.

The computer printer port has 8 data output bits. They are used to select the channel to be checked. There are 4 control output bits normally used for control of the printer. One is used to control the MAX154 and tell it to start conversion. There are also 5 control input bits normally used to receive status signals from the printer. They are now used to allow the computer to read 5 bits of the converter's output. Since it has a 4-bit flash converter, the upper 4 bits are available almost immediately. So, there is no need to check the MAX154's ready line when using Quick BASIC on a slower computer.

The spectrum analyzer option can be built by using the design by Richard Schroeder that appeared in January 1994 *Nuts & Volts* magazine. His "professional-type audio spectrum analyzer" gave the design and values for a 28-channel analyzer. You can add 25 Hz by using a 620K resistor and 20 kHz, with a 680-ohm resistor, to expand his design to 30 channels. You can use every other frequency starting with 25 Hz, for 15 channels. He used 1K and 10 μF as the RC (resistor/capacitor) values for reading peak levels. The signal input to the FETs (Qa, Qb, Qc, and Qd) is the input to the 4051 IC for my design. Richard's design was very good, but it used a lot of LEDs, and of course, using an old computer is cheaper, simpler, and quicker.

Another option is to use any design for a band-pass amplifier such as the ones used in equalizers. Then have the separated frequencies feed the full-wave rectifiers in the audio analyzer.

The project can be built in a small box, about 6 × 8 × 3 inches. Use a larger box if you plan on adding the spectrum analyzer or more channels. Substantial space is needed for the 16 RCA jacks (i.e., 16 inputs). Use a plug-in circuit board for ease of design modification. Plugging and unplugging all 16 inputs is not an easy task. For more than 16 channels, a 44-pin edge connector will not suffice. A solution here is to mount the printer port connector on the board or mount a header connector that plugs into a ribbon cable to the printer port connector. Also, to save on the RCA jacks, you can use a 25-pin "D" type connector where pins 1 to 16 are channels 1 to 16 in, and pins 17 to 25 are grounds.

To connect the inputs to the mixer, use 16 shielded cables about 4 to 6 feet in length. Put RCA plugs on one end and 1/4-inch audio plugs on the other end. For mixers with an "insert" output jack, the 1/4-inch connector needs to be a stereo version. The tip and ring connections are connected together and also connected to the wire going to the converter. On mixers with a "tap" or "output" type of jack, just connect the tip of the 1/4-inch plug (mono or stero) to the wire to the converter. The shield of the wire always connects to the sleeve of the 1/4-inch connector.

Quantity	Number	Source
2	24-pin IC sockets	Newark
4	16-pin IC sockets	Newark
16	14-pin IC sockets	Newark
16	LF353	Newark
4	CD4051	Newark
1	MAX154	Digi-Key
1	circuit board, 4.5 × 6 inches	Newark
1	circuit board socket	Newark
1	metal cabinet, 8 × 6 × 3 inches	Newark
1	LM7805 regulator	Newark
1	LM79L05 regulator	Newark
1	25-pin female connector	Newark
32	RCA jacks	Radio Shack
32	RCA plugs	Radio Shack
32	1/4 stereo plugs	Radio Shack
128 feet	shielded audio cable	Radio Shack
64	1N4148 diodes	Radio Shack
1	18 VCT 1.2-amp transformer	Radio Shack
1	bridge rectifier, 1 amp, 50 V	Radio Shack
1	power cord	Radio Shack
1	on/off switch	Radio Shack
Miscellaneous	resistors and capacitors	Radio Shack

Table 2-18: Materials required

The cable from the audio analyzer that connects to the computer should be kept short (i.e., about 3 to 5 feet). Pull-up resistors of about 470 or 1K on the inputs to the converter should be used to avoid errors caused by noise.

The sensitivity adjustment for most mixers can be set for 0.77 volts. It is a relative setting, however, corresponding to the top of the VU meters. Normally 0.77 volts is about 0 dB, but you can use 0.38 volts for more sensitivity. This voltage is measured on the MAX154 pin 14.

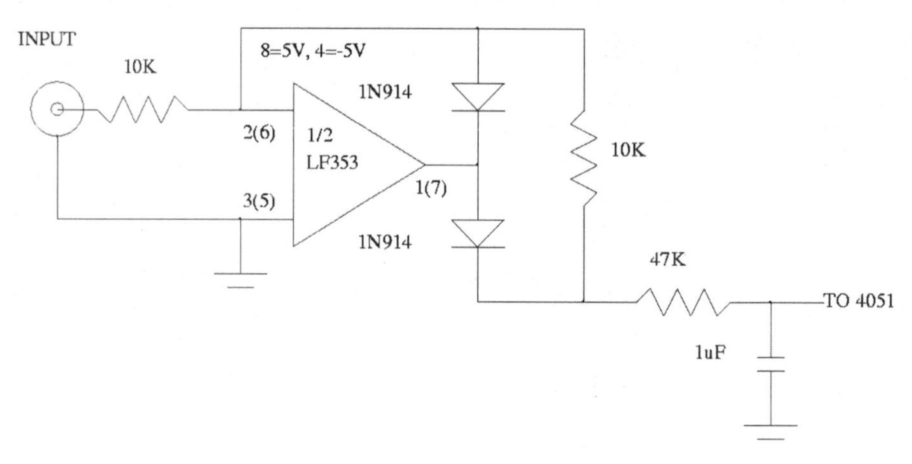

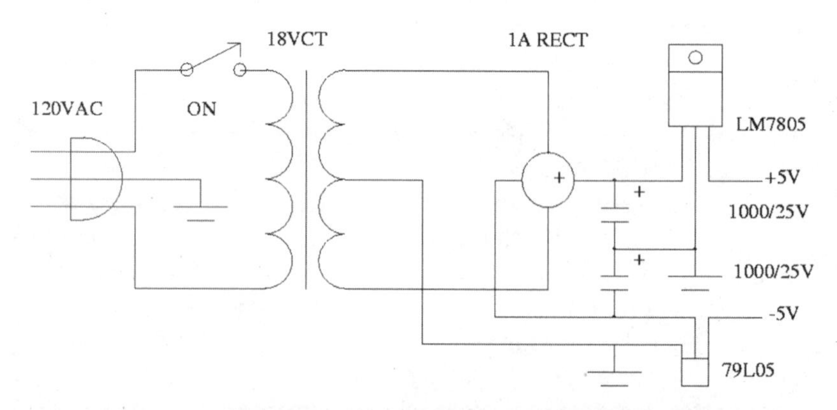

Figure 2-22: Printer port audio analyzer rectifier and power supply

The software listings are for Quick BASIC. A filled box is drawn on the screen for each channel. The size is proportional to the level input for that channel. For a true "VU meter," a formula would have to be added to the software in order to convert the linear result from the analog-to-digital converter to a logarithmic

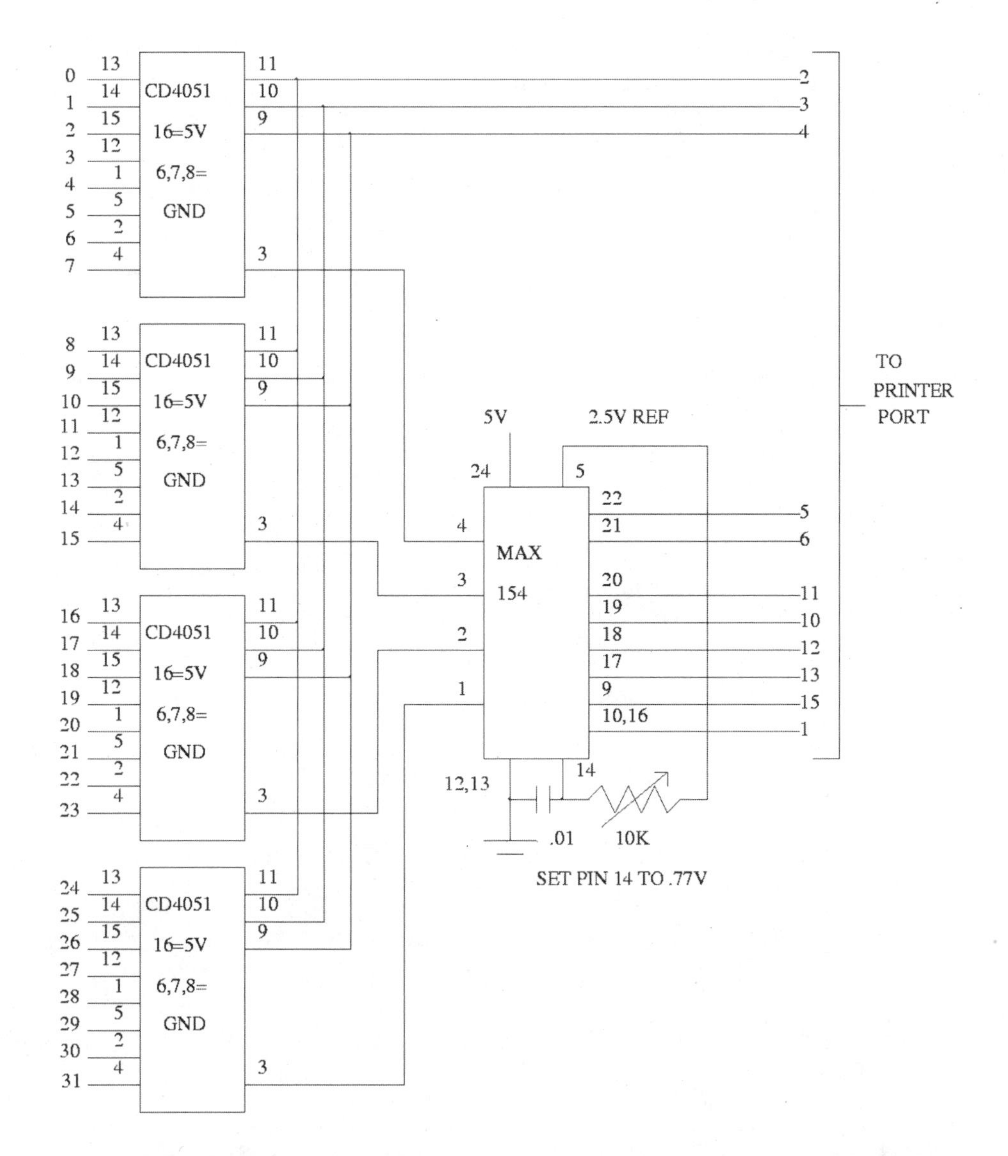

Figure 2-23: Printer port audio analyzer multiplexer and A-to-D converter

value. The display does, however, successfully cover a wide range of audio levels. There is a 16- and a 32-channel version of the software.

The 16-channel version uses a group of "data" tables to create custom characters that will align properly under each VU meter. The 32-channel version avoids this approach, as the space available is even more limited.

```
' 16 CHANNEL PRINTER PORT AUDIO ANALYZER  4/23/96
' BY BOB DAVIS 640 X 200 CGA VERSION
CLS : LOCATE 8, 20: PRINT "PRINTER PORT AUDIO ANALYZER"
LOCATE 11, 20: PRINT "COPYRIGHT 1995 BY ROBERT J DAVIS"
LOCATE 14, 20: INPUT "USE PRINTER PORT NUMBER: ", LPT
DOUT = &H378: COUT = &H37A: CIN = &H379
IF LPT = 1 THEN DOUT = &H3BC: COUT = &H3BE: CIN = &H3BD
IF LPT = 3 THEN DOUT = &H278: COUT = &H27A: CIN = &H279
CLS : SCREEN 1: COLOR 0, 0
FOR A = 10 TO 170 STEP 10                    ' draw lines
LINE (1, A)-(400, A)
NEXT A
FOR B = 10 TO 310 STEP 20                    ' blank for meters
LINE (B - 7, 10)-(B + 7, 170), 0, BF
NEXT B
DATA 0,0,1,0,0,0,1,1,0,0,0,0,1,0,0,0,0,1,0,0,0,0,1,0,0,0,0,1,0,0, 0,1,1,1,0
DATA 1,1,1,1,0,0,0,0,0,1,0,0,0,0,1,0,1,1,1,0,1,0,0,0,0,1,0,0,0,0, 1,1,1,1,1
DATA 1,1,1,1,0,0,0,0,0,1,0,0,0,0,1,1,1,1,1,0,0,0,0,0,1,0,0,0,0,1, 1,1,1,1,0
DATA 1,0,0,0,1,1,0,0,0,1,1,0,0,0,1,1,1,1,1,1,0,0,0,0,1,0,0,0,0,1, 0,0,0,0,1
DATA 1,1,1,1,1,1,0,0,0,0,1,0,0,0,0,1,1,1,1,0,0,0,0,0,1,0,0,0,0,1, 1,1,1,1,0
DATA 0,1,1,1,0,1,0,0,0,0,1,0,0,0,0,1,1,1,1,0,1,0,0,0,1,1,0,0,0,1, 0,1,1,1,0
DATA 1,1,1,1,1,0,0,0,0,1,0,0,0,1,0,0,0,0,1,0,0,0,1,0,0,0,0,1,0,0, 0,0,1,0,0
DATA 0,1,1,1,0,1,0,0,0,1,1,0,0,0,1,0,1,1,1,0,1,0,0,0,1,1,0,0,0,1, 0,1,1,1,0
DATA 0,1,1,1,1,1,0,0,0,1,1,0,0,0,1,0,1,1,1,1,0,0,0,0,1,0,0,0,0,1, 0,0,0,0,1
DATA 0,1,0,0,0,1,1,1,0,1,1,0,0,1,0,0,0,1,0,1,0,0,1,0,0,0,1
DATA 0,1,0,0,1,0,0,0,1,0,1,0,0,1,0,0,0,1,0,1,0,0,1,0,0,0,1,1,1,1,
   0,0,1,1,1,0
DATA 0,1,0,0,0,0,1,0,0,1,1,0,0,0,1,1,0,0,0,1,0,0,0,0,1,0,0
DATA 0,1,0,0,0,0,1,0,0,0,1,0,0,0,0,1,0,0,0,1,0,0,0,0,1,0,0,1,1,1,
   0,0,1,1,1,0
DATA 0,1,0,0,1,1,1,1,0,1,1,0,0,0,0,0,0,1,0,1,0,0,0,0,0,0,1
DATA 0,1,0,0,0,1,1,1,0,0,1,0,0,1,0,0,0,0,0,1,0,0,1,0,0,0,0,1,1,1,
   0,1,1,1,1,1
DATA 0,1,0,0,1,1,1,1,0,1,1,0,0,0,0,0,0,1,0,1,0,0,0,0,0,0,1
DATA 0,1,0,0,0,1,1,1,0,0,1,0,0,0,0,0,0,1,0,1,0,0,0,0,0,0,1,1,1,1,
   0,1,1,1,1,0
DATA 0,1,0,0,1,0,0,0,1,1,1,0,0,1,0,0,0,1,0,1,0,0,1,0,0,0,1
DATA 0,1,0,0,1,1,1,1,1,0,1,0,0,0,0,0,0,1,0,1,0,0,0,0,0,0,1,1,1,1,
   0,0,0,0,0,1
DATA 0,1,0,0,1,1,1,1,1,1,1,0,0,1,0,0,0,0,0,1,0,0,1,0,0,0,0
DATA 0,1,0,0,1,1,1,1,0,0,1,0,0,0,0,0,0,1,0,1,0,0,0,0,0,0,1,1,1,1,
   0,1,1,1,1,0
```

```
DATA 0,1,0,0,0,1,1,1,0,1,1,0,0,1,0,0,0,0,0,1,0,0,1,0,0,0,0
DATA 0,1,0,0,1,1,1,1,0,0,1,0,0,1,0,0,0,1,0,1,0,0,1,0,0,0,1,1,1,1,
  0,0,1,1,1,0
FOR C = 0 TO 15
FOR A = 175 TO 181
   IF C < 9 THEN
      FOR B = 8 + C * 20 TO 12 + C * 20: READ D: PSET (B, A), D: NEXT B
   ELSE
      FOR B = 6 + C * 20 TO 14 + C * 20: READ D: PSET (B, A), D: NEXT B
   END IF
NEXT A
NEXT C
LOCATE 24, 8: PRINT "PRINTER PORT AUDIO ANALYZER";
START:
FOR B = 10 TO 310 STEP 20                 ' DRAW VU METERS
OUT DOUT, (B - 10) / 20                 ' SELECT A:D CONVERTER
OUT COUT, 0: OUT COUT, 255              ' START A:D CONVERSION
LEVEL = (INP(CIN) XOR &H70) \ 8         ' INVERT BIT 7 & SHIFT DATA
LINE (B - 2, 10)-(B + 2, 165), 0, BF' BLANK OLD LINE
LINE (B - 2, (LEVEL) * 5 + 10)-(B + 2, 172), 2, BF
'   LINE (B - 2, (32 - LEVEL) * 5 + 10)-(B + 2, 172), 2, BF
NEXT B
IF INKEY$ = "Q" THEN END
GOTO START
```

```
' 32 CHANNEL PRINTER PORT AUDIO ANALIZER  4/23/96
' BY BOB DAVIS 640 X 200 CGA RESOLUTION
CLS : LOCATE 8, 20: PRINT "PRINTER PORT AUDIO ANALYZER"
LOCATE 11, 20: PRINT "COPYRIGHT 1995 BY ROBERT J DAVIS"
LOCATE 14, 20: INPUT "USE PRINTER PORT NUMBER: ", LPT
DOUT = &H378: COUT = &H37A: CIN = &H379
IF LPT = 1 THEN DOUT = &H3BC: COUT = &H3BE: CIN = &H3BD
IF LPT = 3 THEN DOUT = &H278: COUT = &H27A: CIN = &H279
CLS : SCREEN 1: COLOR 0, 0
FOR A = 10 TO 170 STEP 10                 ' draw lines
LINE (1, A)-(400, A)
NEXT A
LOCATE 23, 2: PRINT "1—CHANNELS—-1625—-FREQUENCY—-16";
LOCATE 24, 8: PRINT "PRINTER PORT AUDIO ANALYZER";
START:
FOR B = 10 TO 310 STEP 10                 ' DRAW VU METERS
OUT DOUT, (B - 10) / 10                 ' SELECT A:D CONVERTER
OUT COUT, 0: OUT COUT, 255              ' START A:D CONVERSION
LEVEL = (INP(CIN) XOR &H70) \ 8         ' INVERT BIT 7 & SHIFT DATA
LINE (B - 2, 10)-(B + 2, 165), 0, BF' BLANK OLD LINE
LINE (B - 2, (LEVEL) * 5 + 10)-(B + 2, 172), 2, BF
NEXT B
IF INKEY$ = "Q" THEN END
GOTO START
```

Printer Port Energy Management 2-5

This circuit could be a replacement for the 8052-based version found elsewhere in this book. IBM PCs have become so cheap that I've picked them up along the road when they were being thrown out with the garbage. Then, of course, they can be used for projects such as printer port devices. I felt it was time to dedicate a PC to the job of energy management. The monitor could be turned off, and a hard drive is not needed. The computer therefore would consume a relatively small amount of power. However, if the monitor is used, it could provide a lot more than a three-digit LED display. Information displayed could include all eight input temperatures at one time, history information, and the status of the outputs. The keyboard gives more control as well.

To provide the same functions as the 8052-based circuit, I needed eight analog inputs and eight digital outputs. The printer port could easily provide eight digital outputs. The four control bit outputs and five control inputs are then available to control the analog-to-digital converter and input data. To get 8 data bits through four lines, a 74LS157 2:1 multiplexer is used. The other input bit could be used to monitor the end-of-conversion signal.

To run the ADC808 analog-to-digital converter, a 600-kHz clock is needed. This is obtained from a 10-MHz oscillator and a 74LS393 divide-by-16 circuit, which gives about 625 kHz. Three bits are needed to select the analog input port, and another bit is used to select high or low 4 bits to be sent back to the computer. These are borrowed from the 8-data-bit outputs of the printer port. Since these bits are serving a dual purpose, a latch must be used in order to capture data output that is then used to control devices. This 8-bit latch is not an extra IC—it also serves as a necessary buffer circuit.

Reference voltages of 1.25 and 3.75 volts are used once again. These are based on a stable 5-volt source and a 2.5-volt regulator. The LM335 provides an output voltage that corresponds to temperature in kelvins. The conversion is done like this: 273 kelvins is 0 degrees Celsius or 32 degrees Fahrenheit and 2.73 volts. Fahrenheit is 9/5 times Celsius + 32 degrees. At the low end, 2.43 volts is –30 degrees Celsius or –22 degrees Fahrenheit. At the other end 3.73 volts is 100 degrees Celsius or 212 degrees Fahrenheit. So we will usually be using 2.43 to 3.73 volts, well within the 1.25 and 3.75 limits. The only problem may be at the high end. There, a little more headroom would be helpful when monitoring a boiler or water heater. When monitoring a boiler they sometimes overshoot, but their safety shutoff switch is usually set to under 212 degrees.

Quantity	Number	Source
1	28-pin IC socket	Newark
1	20-pin IC socket	Newark
1	16-pin IC socket	Newark
2	14-pin IC sockets	Newark
1	74Ls393	Newark
1	10-MHz clock	Newark
1	74Ls157	Newark
1	74Ls374	Newark
1	ADC0808	Jameco
1	circuit board, 4.5 × 6 inches	Newark
1	circuit board socket	Newark
1	metal cabinet, 6 × 4 × 2 inches	Newark
1	LM7805 regulator	Newark
1	LM336 2.5-volt reference	Digi-Key
8	LM335 temperature sensors	Digi-Key
1	25-pin female connector	Newark
1	9 VDC AC adapter	Radio Shack
Miscellaneous	resistors and capacitors	Radio Shack

Table 2-19: Materials required

Regarding temperature steps, 256 steps correspond to 2.5 volts, so there is 1 step per degree Celsius. This is 9/5 of a degree Fahrenheit, or about 2 degrees per step. This is usually adequate, but the ADC808 can be replaced with a ADC810 for 10 digits, providing a resolution of 0.5 degrees per step.

```
' PRINTER PORT ENERGY MANAGEMENT 1/23/98 BY BOB DAVIS
CLS
LOCATE 20, 10: PRINT "ENERGY MANAGEMENT, BY BOB DAVIS"
LOCATE 22, 18: INPUT "USE PRINTER PORT NUMBER: ", LPT
DOUT = &H378: COUT = &H37A: CIN = &H379
IF LPT = 1 THEN DOUT = &H3BC: COUT = &H3BE: CIN = &H3BD
IF LPT = 3 THEN DOUT = &H278: COUT = &H27A: CIN = &H279
'LOCATE 18, 1: PRINT "BUILDING 8 ="
START:
FOR B = 0 TO 7
OUT DOUT, B                             ' SELECT A:D CONVERTER
OUT COUT, 0                   ' START A:D CONVERSION (inverted)
OUT COUT, 1
FOR A = 1 TO 20: NEXT A                ' DELAY
level = (INP(CIN) XOR &H80) \ 16      ' INVERT BIT 7 & SHIFT DATA
OUT DOUT, B + &H8                      ' GET NEXT 4 BITS
```

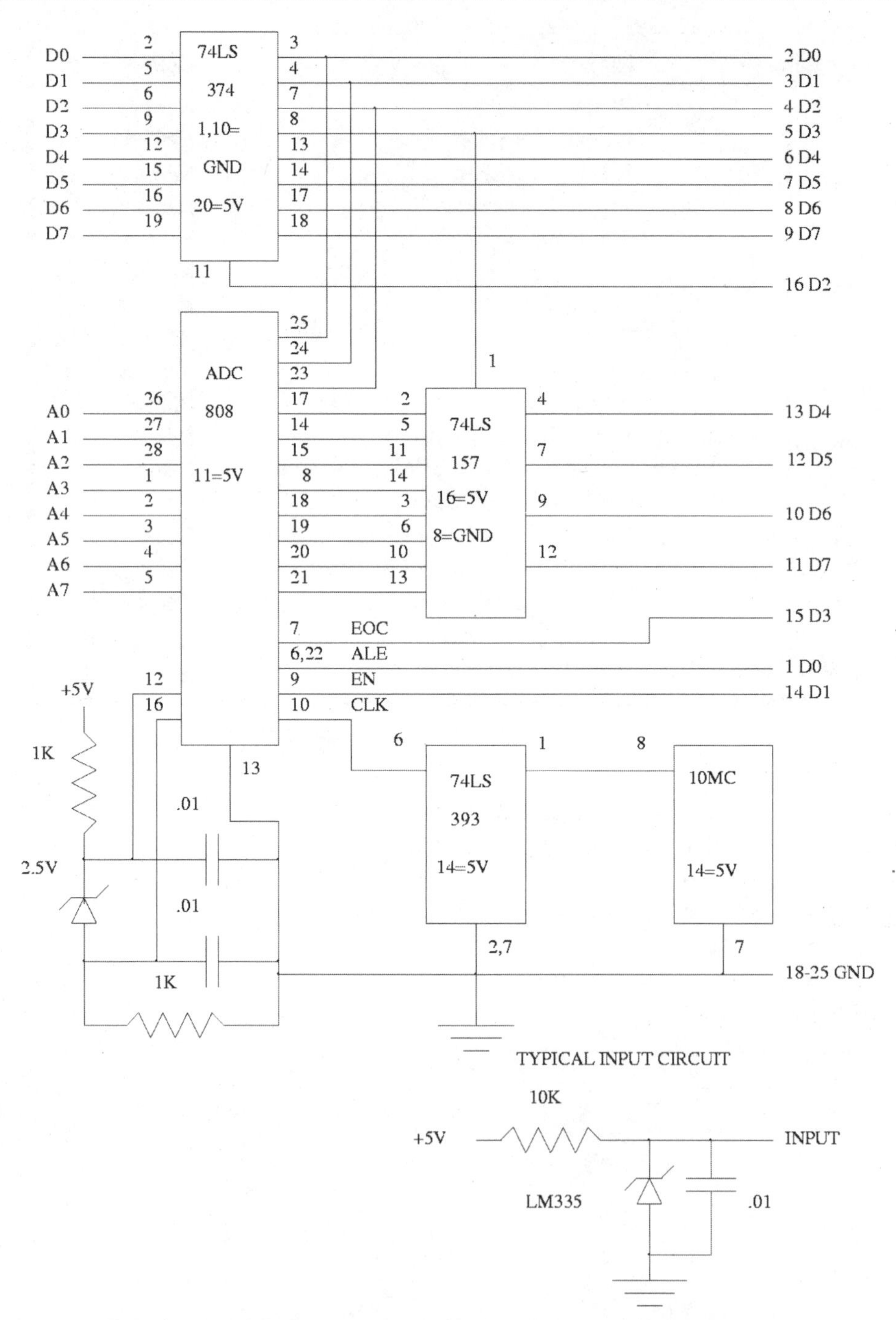

Figure 2-24: Printer port energy management schematic

```
level = level + ((INP(CIN) XOR &H80) \ 16) * 16
LOCATE (B * 2) + 4, 13: PRINT level
' NOTE: TRIM REFERENCE FOR 2.55V
IF level > 0 THEN
   LOCATE (B * 2) + 4, 33: PRINT (level - 158)
   LOCATE (B * 2) + 4, 53: PRINT ((level - 158) * 9 / 5) + 32
END IF
NEXT B
IF INKEY$ = "Q" OR INKEY$ = "q" THEN END
GOTO START
```

Printer Port System Clock 2-6

Many schools and businesses across this country, and around the world, use a synchronized clock system to keep all of their clocks exactly the same. This same system also provides class bells or sounds a horn to signal time to start and stop work. These systems cost nearly $1,000 for just the controller. You could instead use an old IBM PC and simple interface circuitry.

The most common synchronized clock design uses a 24-volt 1-second pulse, once a minute, in order to advance the clocks. Next, there is a two-part correction system. Polarity is reversed to slow down fast clocks a few minutes before each hour. Extra pulses are sent during the reverse polarity time at 1 minute before the hour to speed up slow clocks. This then corrects, every hour, for slow or fast clocks. Correction takes care of clocks that are off several minutes per hour. The correction pulses are about 15 pulses for slow clocks, and 10 for fast clocks per hour. This corrects clocks that are losing minutes because of power interruptions or worn components, or those that double-click and gain time. This system will automatically correct up to a 1-hour power loss, taking 4 hours to complete the correction. Any more time correction than that would require manual intervention.

The oldest clock system design that I know of used a spring-powered pendulum clock. There were contacts in the clock that fed relays to run the synchronized clocks. The 24-volt pulses, in turn, wound the spring to keep the clock running. The spring then could power the master clock for several minutes after power was lost. The pendulum height could be adjusted to speed up or slow down the master clock. These old clock controllers have generally become worn to the point that they are no longer operational.

Over the years, I've made my own versions of controllers. They used logic circuits and static RAM to store times and sound various class bells. The biggest problem with homemade versions is setting the master clock. No matter what you use, it needs periodic adjustment, and at least twice a year it has to be set for daylight saving time. Some circuit designs that I tried needed adjustments each and every month. Since they keep time in binary, the only way to accurately set them was by pressing a reset button, exactly on the hour. Many versions lost their time with power interruptions, and bells rang at the wrong times. All designs had many problems, and it would be impossible to recommend them for regular use. In addition, they were too complex for an inexperienced builder.

Then there are the professional controllers. They use an 8-bit microprocessor controller, static RAM, battery backup, and a host of features. I've seen some that remember when power goes out—then when power is restored, they automatically reset all clocks, even hours later! Some have as many as eight bell circuits so different class schedules can all have their own bell times. An example might be a grade school, a middle school, and a high school all having their own schedules. Another example would be a factory with different shifts in different areas of a factory complex. These professional units usually cost $900 or more, but are very reliable. However, even these are becoming outdated. They are extremely overpriced and are difficult to set for bell times.

Then, there is this homemade version that costs less than peanuts, is easy to set, and can have six bell circuits, if needed. The catch is, it needs an old IBM-compatible computer as its controller. I used a 386SX-16 IBM PS-2 computer. My school was fortunate and had several of them donated to us. Time is easy to set on a computer. All you need to do is just exit the clock program and type "time." Then enter the correct time and restart the program. Advancing the clock, and setting it back, is as easy as pressing a key on the keyboard. Although the included software is quite simple, you can add any feature you want, with your own software additions. A cost-effective software modification could be added to allow the clock controller to remember the last time it had power. When power is lost and then comes back on, the controller would automatically reset the clocks. It could even figure out daylight saving time and adjust the clocks accordingly, with the proper software revisions.

My controller uses the printer port, where bits 0, 1, and 2 drive three relays by way of switching transistors. Relays could be easily added if you need them for additional alarms or bells. Bit 0 turns on every minute, for 1 second, to advance the clocks. Bit 1 activates 10 minutes before the hour to reverse the pulse polarity, for any fast clock. Then bit 0, at 1 minute before the hour, starts pulsing every few seconds to provide 18 correction pulses per hour for clocks that are slow or have lost power. Bit 2 pulses a 12 VAC buzzer mounted on the back of the clock to signal the start of and to dismiss all of the classes. You could use 12 or 24 volts DC for other kinds of bells or buzzers.

When installing this system, I did encounter a couple of problems. There were three separate buildings on our campus that required synchronized clocks. They were originally wired using 18-gauge thermostat-type wire. Over the years, those cables were replaced with plain old telephone wire. The interface unit delivered 30 volts under no load, but under a load (i.e., when advancing the clocks), it only delivered about 25 volts. This provided only about 22 volts, or less, to the last clock. This power loss was due to the line resistance (i.e., 500 feet of 24-gauge telephone wire). Two of the clocks would not operate with such a low voltage. They were quickly "revived," however, by using two telephone lines in parallel to reduce the voltage drop. Surprisingly, three clocks at another building more than 400 feet away had no problem with the new controller.

Another computer problem occurs when power is lost and then restored. The computer tests the parallel port both in CMOS and when DOS loads. These tests will advance the clocks and sound the buzzers repeatedly. A UPS (uninterruptible power supply) will remedy this situation for short power outages. I do not know of a solution for long-term outages. Perhaps a delay circuit that would keep the controller from coming on line until a minute after power is restored would solve this problem.

For future consideration, the program could save the time on its hard drive and update it every minute. Then, when power is restored, it could check this stored

Qty	Part
3	relays, 12 VDC 3-amp contacts, part # 275-206 (plug in) or 275-249 (solder in)
3	2N2222 transistors
3	1N4001 (or equivalent) diodes
3	1 K resistors
1	6 × 10 × 2 inch box
1	4 × 4 inch circuit board with edge connector
1	jack for edge connector
1	25-pin male plug and hood to match
1	24-volt 2-amp (minimum) center tapped power transformer
1	4-amp, 50-volt bridge rectifier
1	2200-µF (or 4,00 µF) 50-volt capacitor
2	220-µF, 25-volt capacitors

Table 2-20: Parts list (Most parts are available at Radio Shack)

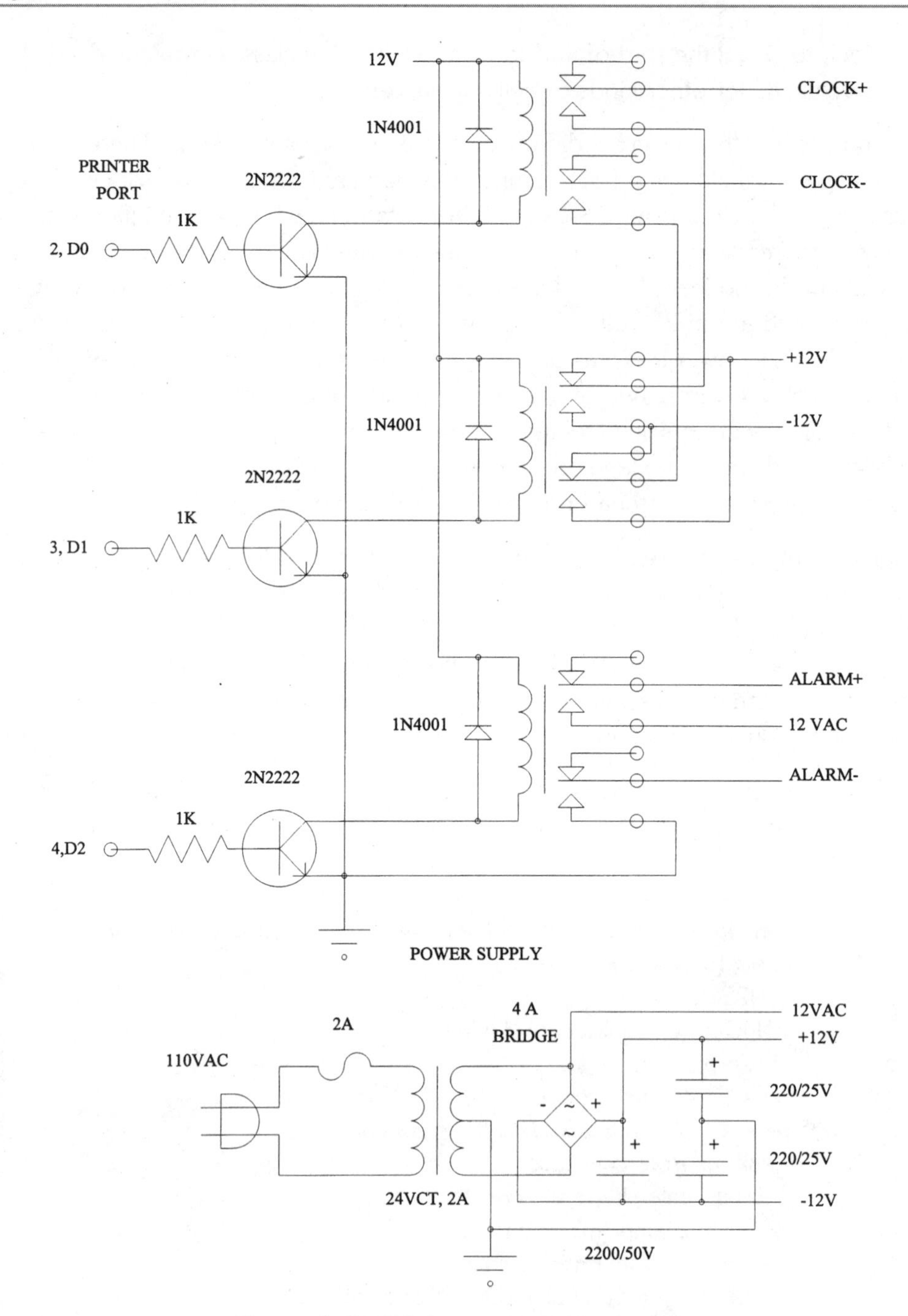

Figure 2-25: Printer port clock circuit

time and the current time, and then automatically reset all clocks at high speed. Another problem I've encountered is that when computers are left on for long periods of time, their CMOS clocks tend to lose the correct time. Some computers, after long power outages that exceed their UPSs, can come back on with their clocks more than a day off! The solution there might be to add a trimmer to the CMOS clock, or use an old 286, as they had trimmers to allow adjustment of their clocks. Another idea might be to use software to interface the computer to the atomic clock in Washington. This would automatically correct the time every day, or every time power is restored.

```
CLS : LOCATE 1, 1
PRINT "Computer controlled clock by Bob Davis "
INPUT "Use printer port number: ", LPT
PRINT ""
PRINT "Press 'S' to stop clocks, 'F' to Fast forward clocks"
PRINT "Press 'Q' to Quit this program, 'R' to Run normally"
Dout = &H378
IF LPT = 1 THEN Dout = &H3BC
IF LPT = 3 THEN Dout = &H278
LOCATE 7, 1: PRINT "Time      Advance     Polarity      Bell     Command"
LOCATE 8, 1: PRINT "—        ———       ————         ——       ———"
Start:
keypress$ = INKEY$
IF keypress$ = "R" THEN Normal = 0: Fast = 0
IF keypress$ = "Q" THEN STOP
IF keypress$ = "S" THEN Normal = 1
IF keypress$ = "F" THEN Fast = 1
hrsmin$ = MID$(TIME$, 1, 5)
hours = VAL(MID$(TIME$, 1, 2))
minutes = VAL(MID$(TIME$, 4, 2))
Seconds = VAL(MID$(TIME$, 7, 2))
Tenths = VAL(MID$(TIME$, 8, 1))
IF minutes <> Oldmin THEN Oldmin = minutes
IF Normal = 0 AND Seconds = 0 THEN advance = 1 ELSE advance = 0
IF (Normal = 0 AND minutes = 59) OR Fast = 1 THEN
IF Tenths = 3 OR Tenths = 6 OR Tenths = 9 THEN advance = 1 ELSE advance = 0
END IF
IF Hours = 8 OR Hours = 12 OR Hours = 13 OR Hours = 17 THEN
IF minutes = 0 AND Seconds = 0 THEN bell = 1 ELSE bell = 0
END IF
IF minutes > 51 THEN polarity = 1 ELSE polarity = 0
LOCATE 9, 1: PRINT hrsmin$
LOCATE 9, 12: PRINT advance
LOCATE 9, 23: PRINT polarity
LOCATE 9, 34: PRINT bell
LOCATE 9, 45: PRINT keypress$
OUT Dout, advance + (polarity * 2) + (bell * 4)
GOTO Start
```

Printer Port Line Status Monitor 2-7

When you have 12 or more telephone lines coming into your telephone system, it is possible for one or more of them to stop working without your even knowing it. That is, until someone calls up and says they get a busy signal when they dial "9" to get an outside line. Or worse yet, when someone calls and says they've been calling all week and no one answered. Then you have to dig out a voltmeter and check the lines to see if they are dead, working, stuck, or not in use. This quite often involves tapping into the lines and disrupting telephone calls in progress. If there were only some easy way to determine how long it has been since the line was last available. You would then know for sure if a low-voltage reading was someone on the line or a stuck line.

Another problem with a stuck telephone line is that the local telephone company, in our area, will not give a "flat rate" for local calls to businesses in this area. We are charged a whopping five cents a minute for every call leaving the campus. Even if our long distance carrier is only charging 10 cents a minute, there is an additional charge of 5 cents a minute to reach the long distance company! This is added to the charge per line just to have the line working. Our local phone bill usually exceeds $500 a month, and sometimes exceeds $700! I keep

16	4N30 opti-couplers
16	8-pin sockets
16	100K resistors
16	15K resistors
2	74LS153 ICs
2	16-pin sockets
5 feet	nine-conductor wire
1	25-pin plug and hood to match
1	4-inch circuit board with edge connector
1	jack for edge connector
1	box, 6 × 10 × 2 inches
1	9 VDC AC adapter with a 7805 voltage regulator
Miscellaneous	resistors and capacitors

Table 2-21: Parts list

dreaming of someone coming along with a "direct to satellite" local telephone service that would bypass these local phone companies.

That is where my telephone line status monitor comes in. It can monitor up to 32 telephone lines, essentially giving a "busy line" display at a fraction of the cost of professional units. Once again, the only catch is that you need a computer—but we have numerous donated 386 SX-16s, so let's put another one to use. In actuality, I combined this with the clock circuit and software so one computer serves both purposes.

On a telephone line, there are three possible and measurable voltages. If the line is good, but no call is in progress, then there should be 48 to 52 volts on the line. If a call is in progress, there should be 8 to 12 volts on the line. If the line is dead or disconnected, it should read 0 volts. Any device that's connected to the line must exceed 100 kohm, and therefore will not load down the line below the 48-volt threshold. Most opti-couplers would not work, since they aren't sensitive enough to meet the 100K resistance requirement. This project requires a more sensitive opti-coupler, one with a Darlington transistor in its output. At the same time the opti-coupler must not be too sensitive as to where it does not turn off when the voltage drops to 8 to 12 volts. I used the 4N30, although others might work, but perhaps with different resistor values.

The schematic shows 8 of 16 circuits, but by changing the multiplexer to an 8-to-1 device you could monitor up to 32 telephone lines. This circuit *does not work on digital telephone lines*, only on analog telephone lines. Digital telephone lines require reading the switchboard's computer to tell if they are in use—usually an old digital telephone will do for that purpose. Perhaps in the near future I will develop an interface to a computer that will allow it to simulate a busy line display type of digital telephone.

Not shown in the schematic, Figure 2-26, are pull-up resistors on pins 2 and 14 of the 74LS153. They should be about 220- to 470-ohm and go to the 5-volt source. This is once again to reduce noise on the cable to the printer port. The circuit works fine without them but may not work with some computers.

In the software listing you can substitute your telephone exchanges for the three question marks shown as "(???)."

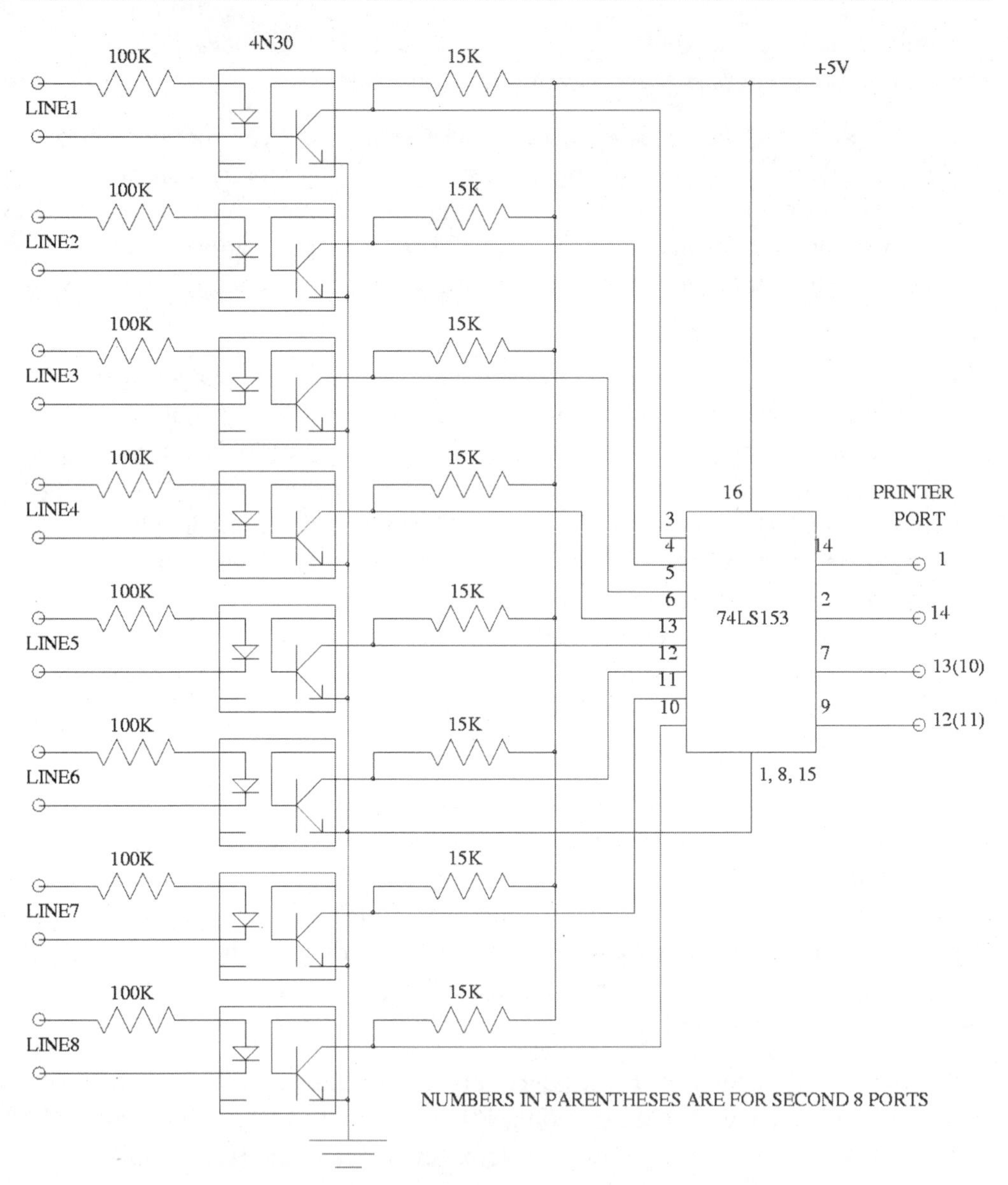

Figure 2-26: Printer port telephone line monitor

```
CLS :LOCATE 1, 1
PRINT "Computer controlled telephone line monitor by Bob Davis "
INPUT "Use printer port number: ", LPT
PRINT ""
PRINT "Press 'Q' to Quit this program, 'R' to Run normally"
dout = &H378: cout = &H37A: cin = &H379
IF LPT = 1 THEN dout = &H3BC: cout = &H3BE: cin = &H3BD
IF LPT = 3 THEN dout = &H278: cout = &H27A: cin = &H279
LOCATE 13, 10: PRINT " (???)       (???)       (???)       (???)"
LOCATE 14, 10: PRINT " ---         ---         ---         ---"
PRINT "line 1"
PRINT "line 2"
PRINT "line 3"
PRINT "line 4"
FOR B = 1 TO 4                                     ' Line monitor
OUT cout, (B - 1)                              ' SELECT MUX
status = INP(cin)
LOCATE 14 + B, 11
IF (status AND &H10) = 0 THEN
  LINETIME(1, B) = 0: PRINT "OPEN "
ELSE
  IF LINETIME(1, B) = 0 THEN LINETIME(1, B) = 1: PRINT hrsmin$
END IF
LOCATE 14 + B, 22
IF (status AND &H20) = 0 THEN
  LINETIME(2, B) = 0: PRINT "OPEN "
ELSE
  IF LINETIME(2, B) = 0 THEN LINETIME(2, B) = 1: PRINT hrsmin$
END IF
LOCATE 14 + B, 33
IF (status AND &H40) = 0 THEN
  LINETIME(3, B) = 0: PRINT "OPEN "
ELSE
  IF LINETIME(3, B) = 0 THEN LINETIME(3, B) = 1: PRINT hrsmin$
END IF
LOCATE 14 + B, 44
IF (status AND &H80) <> 0 THEN
  LINETIME(4, B) = 0: PRINT "OPEN "
ELSE
  IF LINETIME(4, B) = 0 THEN LINETIME(4, B) = 1: PRINT hrsmin$
END IF
NEXT B
GOTO Start
```

Printer Port Audio Mixer 2-8

For this project you will need to sit down and use your imagination. Think of a sound system where there is no longer a “snake” cable to run from the stage to the mixer. Think of a sound system that doesn’t have a mixer covered with mechanical controls—controls that wear with use and that need periodic cleaning. Imagine a sound system where all settings can be arranged with a few keystrokes, and where all of your adjustments can be saved during rehearsals, then recalled later during the performance. Imagine a computer that could even be trained to run the sound system for you.

Quantity	Number	Source
19	UPC1406	MCM Electronics
8	LF351	MCM Electronics
8	TDA1524	MCM Electronics
6	LF353	MCM Electronics
9	MAX528	Digi-Key
1	MAX158	Digi-Key
1	7HCT138	Mouser
1	LM7912 voltage regulator	
1	LM7812 voltage regulator	
1	LM79L05 voltage regulator	
1	LM78L05 voltage regulator	
12	1N914 diodes	
14	8-pin sockets	
8	18-pin sockets	
10	16-pin sockets	
9	20-pin sockets	
1	28-pin socket	
8	XLR input jacks	
6	1/4-inch output jacks	
1	power cord	
1	on/off switch	
1	25 VCT, 1-A power transformer	
1	1-amp, 50-volt bridge rectifier	
2	2200 µF/16-volt capacitors	
5 feet	20-conductor wire	
1	25-pin plug and hood to match	
1	circuit board, 4 × 4 inches, with edge connector	
1	jack for edge connector	
1	box, 6 × 10 × 2 inches	
Miscellaneous	resistors and capacitors	

Table 2-22: Parts list

Well, imagine no more—it's almost here! The only catches are that you have to build it yourself, because no one makes our "dream system" yet, and that the quality and number of channels will be limited. It will in fact cost more to build than to buy a conventional mixer.

The "computer controlled audio mixer" contains no variable resistors or other conventional controls. This mixer is entirely controlled by a computer. All controls and displays (levels, etc.) are on the computer screen. All controls are accessed through the computer keyboard and setup information could be stored in the computer storage devices. Theoretically, the computer could run the mixer, to a degree, without human intervention. At the operator's request, the computer display could be changed to show other operating parameters, including such things as an "oscilloscope-type" display of any input or output of the mixer. Software for all of these additional features hasn't been developed yet.

This mixer is not an extremely high-quality device because of the lack of high-quality ICs necessary for this purpose. Perhaps this project will encourage some manufacturers to produce the needed chips. It does work quite nicely for demonstration purposes or in low-key applications such as churches or conferences. Distortion is around 0.5% and if you turn up all the controls, clipping from overdriven amplifiers can be a problem because of the somewhat limited dynamic range.

The idea of a computer-controlled mixer came to me one day as I was contemplating rebuilding a mixer. The controls get worn and noisy and typically need cleaning or replacement. The idea of computer control eliminates another common problem. Some people lack the skills needed to set up a mixer, and others might have preferences that are different from yours. You would no longer have to reset all the controls to their original settings if someone like that used the mixer.

The first problem was finding ICs to do the job. I'm not aware of anyone that makes the needed ICs. If they do, they aren't available through common sources. Then, I discovered the TDA1524 in an old manual. After looking through several catalogs I found it in an MCM Electronics catalog. At about that time, Maxim was advertising demos of its new 4- and 8-channel digital-to-analog and analog-to-digital converters. Combine those devices and there you have it. What I would like to see is an all-in-one IC with eight digital-to-analog converters, gain, volume, balance, bass, midrange, treble, send 1, and send 2. Even better would be to have high midrange, low midrange, and two more sends, or about 12 controls per channel! Most importantly, I would like better signal-to-noise figures, less distortion, and more dynamic range.

The TDA1524 has its drawbacks. There is no midrange tone control and the specifications aren't too impressive. It does work well enough to demonstrate the concept. The TDA1524 was designed for car stereos and uses 9 to 18 volts as its power source. This IC supplies a 4-volt reference voltage that feeds four variable resistors. In normal operation, the variable resistors via the wipers return a voltage of 0 to 4 volts to the treble, bass, volume, and balance inputs. The voltage then performs the desired function within the IC. If the reference voltage was instead fed to an analog-to-digital converter, the converter could then supply the 0 to 4 volts to control the IC.

Supplying 0 to 4 volts causes problems for 5-volt-powered analog-to-digital converters. They can't swing their outputs high enough. The MAX510 can swing "rail to rail" but it has only four converters. My desire was to have at least eight controls per channel, which would require a converter with eight outputs. The MAX529 has eight outputs but can't provide more than 2.75 volts out. That leaves only the MAX528 running with something below –3 volts and a little more than +8 volts in order to deliver the needed 0- to 5-volt range. This setup is actually ideal. The TDA1524 needs +9 to +18 volts. So a power supply voltage of +9 and –9 volts agreeable to both ICs. The MAX528 has a maximum differential power supply of 20 volts, which also works fine at these voltages. The reference inputs will be ground and the 4 volts from the TDA1524.

Then there are four other controls needed for each channel: Monitors, Effects, Cueing, and Sends. In this case using a TDA1524 for each would be a waste. One IC containing four voltage-controlled amplifiers was needed but is not easily obtained. Many companies make one or two voltage-controlled amplifiers per chip—and then there is the UPC1406, also available from MCM Electronics. It has two gain controls per IC in a nine-pin in line package. That means that two of them fit nicely into an 18-pin socket. It does, however, have its drawbacks.

The first drawback is that the UPC1406 needs a 0- to 6-volt control voltage. That can be accomplished by dividing the MAX528 into two four-channel digital-to-analog converters. The first four drive the TDA1524 and use its 4-volts reference. The next four drive the UPC1406s and use its 6-volt reference.

The second problem is that the UPC1406 has no gain and a much lower output level than the TDA1524, but that can be compensated for with resistors in the summing amplifier circuit.

The MAX528 is available from Digi-Key. Maxim provides a detailed specification sheet, and the operation is much easier than it appears. To program a voltage, you first bring the chip select low, then you send serial data and clock pulses. There are 16 bits to send. The first eight are address bits that select which of the eight converters you are programming. All eight can be set at the same time. Then, there are eight data bits that correspond to a value of 0 to 255, representing minimum to maximum volume. After the 16 bits are sent, raise the chip select high and the MAX528 will output the new control voltages. Its really that simple to program and control.

The next circuit is the summing amplifiers and master controls. The summing amplifiers add inputs from eight input channels and give an output that represents the sum of the eight inputs. By changing the gain control resistor from 10K to 22K, the summing amplifiers for these four send channels can have some gain as well. The master controls also use the MAX528 and UPC1406. Here, there can be an additional problem or two. The UPC1406s reference output can only drive four of the MAX528s digital-to-analog converters. The MAX528 needs a reference input from two of the UPC1406s. In addition, the UPC1406 has a limited dynamic range. It clips if the input exceeds 2 volts peak-to-peak. Take care not to overdrive it. Reducing the gain in the summing amplifiers will help reduce this clipping.

To allow the computer to display output signal levels, the MAX158 an eight-channel analog-to-digital converter is used. A MAX154 could be used to monitor four inputs, on a smaller mixer, with only a few software changes. It is quite easy to use—just provide inputs to convert, select the analog channel, and read the results. It has a 2.5-volt reference and a 4.7K resistor drops that voltage to 0.5 volt.

The send, queuing, monitor, and effects are referred to as send 1 to send 4. You can designate your own order and names for these controls in the software. To save a few dollars, you can omit them if you don't need their outputs for your mixer.

The typical printer port has 8 bits of data output. These are used to select the channel to be controlled. A 74HCT138 selector decodes D0, D1, and D2 in order to select one of eight channels. Bits D3 and D4 are available for future expansion of up to 32 channels. The other bits, D5 and D6, select the master gain controls and the analog-to-digital converters.

As an application, this mixer could be found on a stage. It would appear as a box with microphone jacks on it. A small control cable would go to the computer, which could be remotely located anywhere in the building. Also, it is possible for a transmitter to be used in order to eliminate the need for any connecting cable. There are transmitters that are advertised for remote printer operation; however, this mixer has not been tested with one, and so I can't guarantee successful operation.

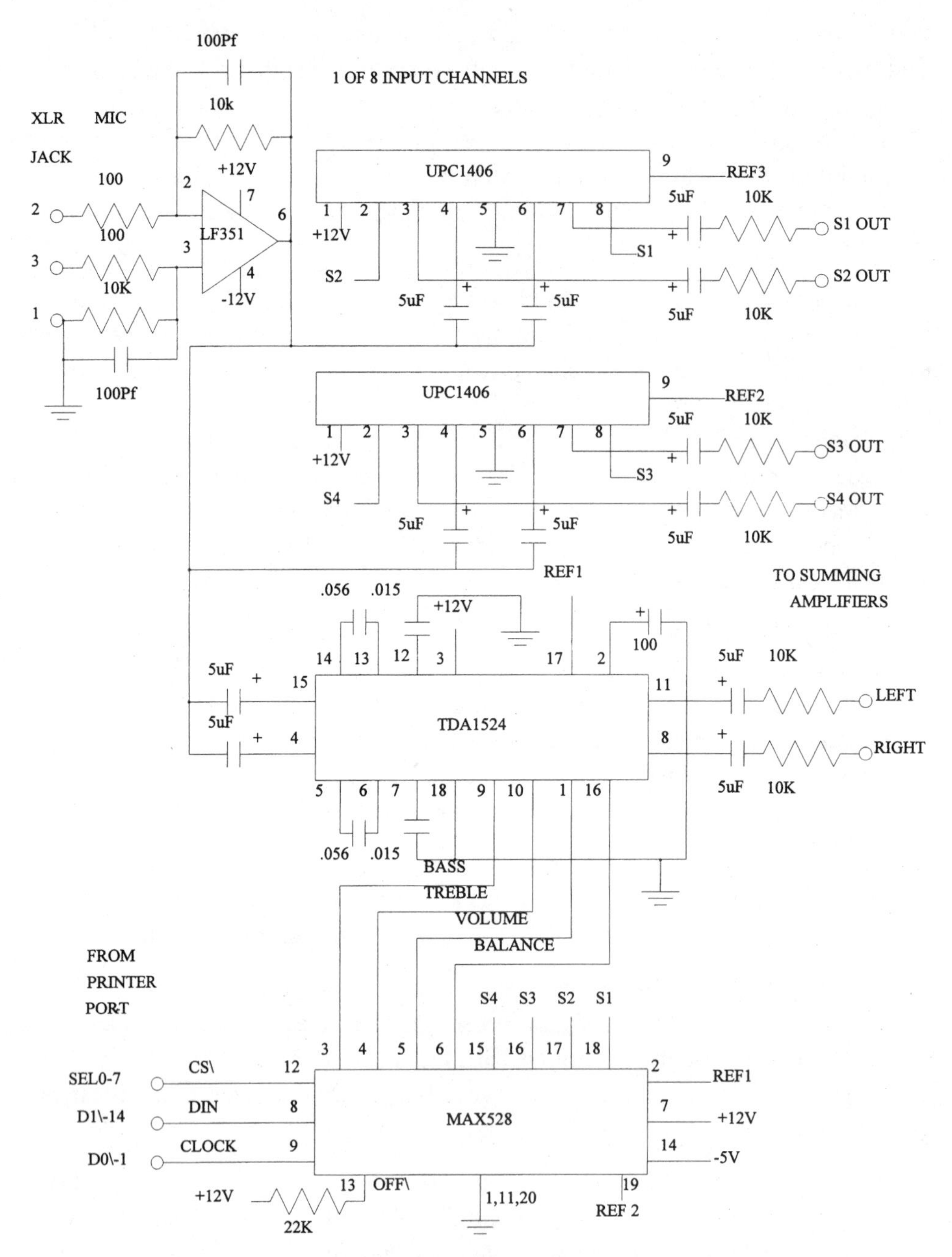

Figure 2-27: Printer port audio mixer input channels

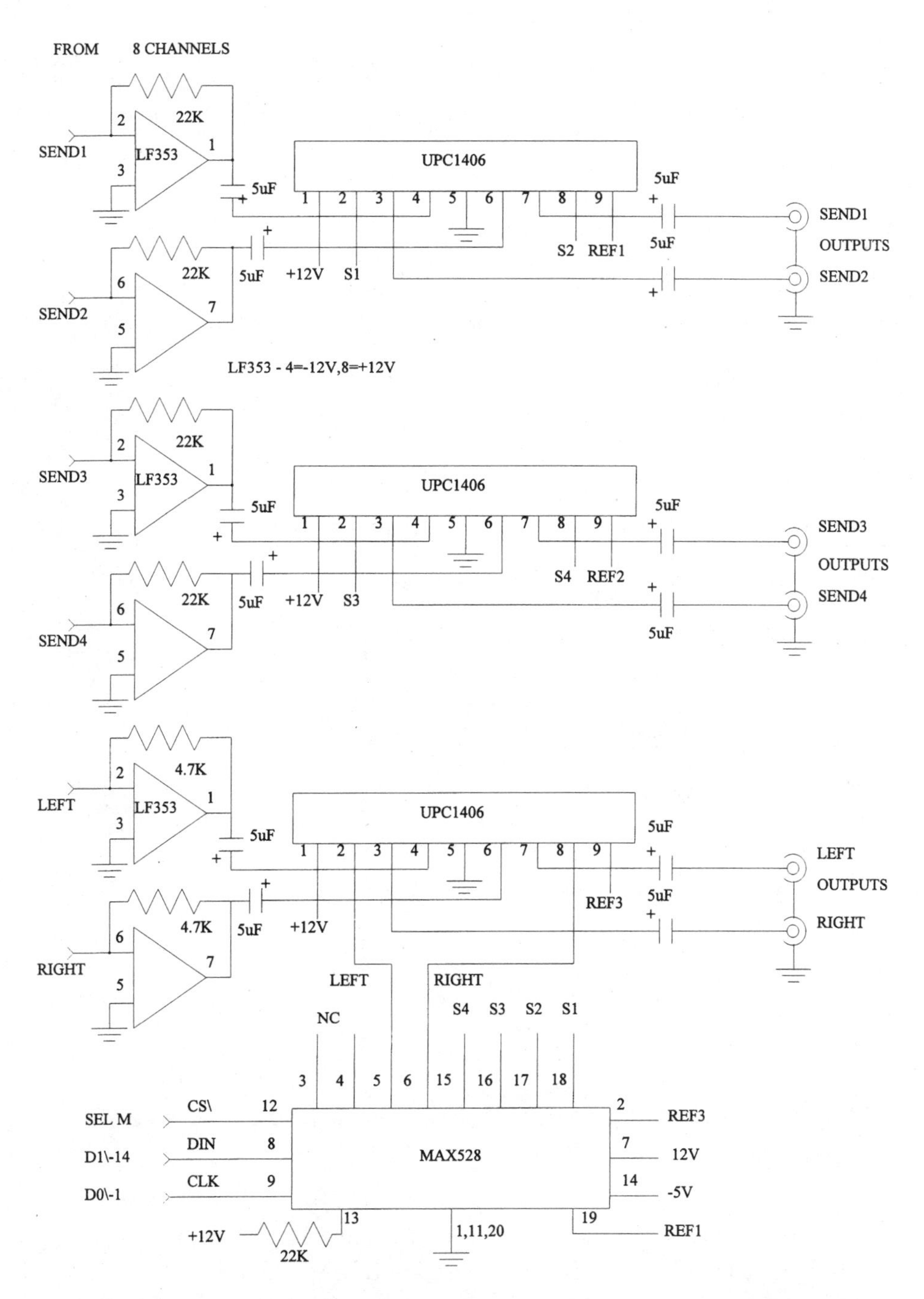

Figure 2-28: Printer port audio mixer summing and master amplifier

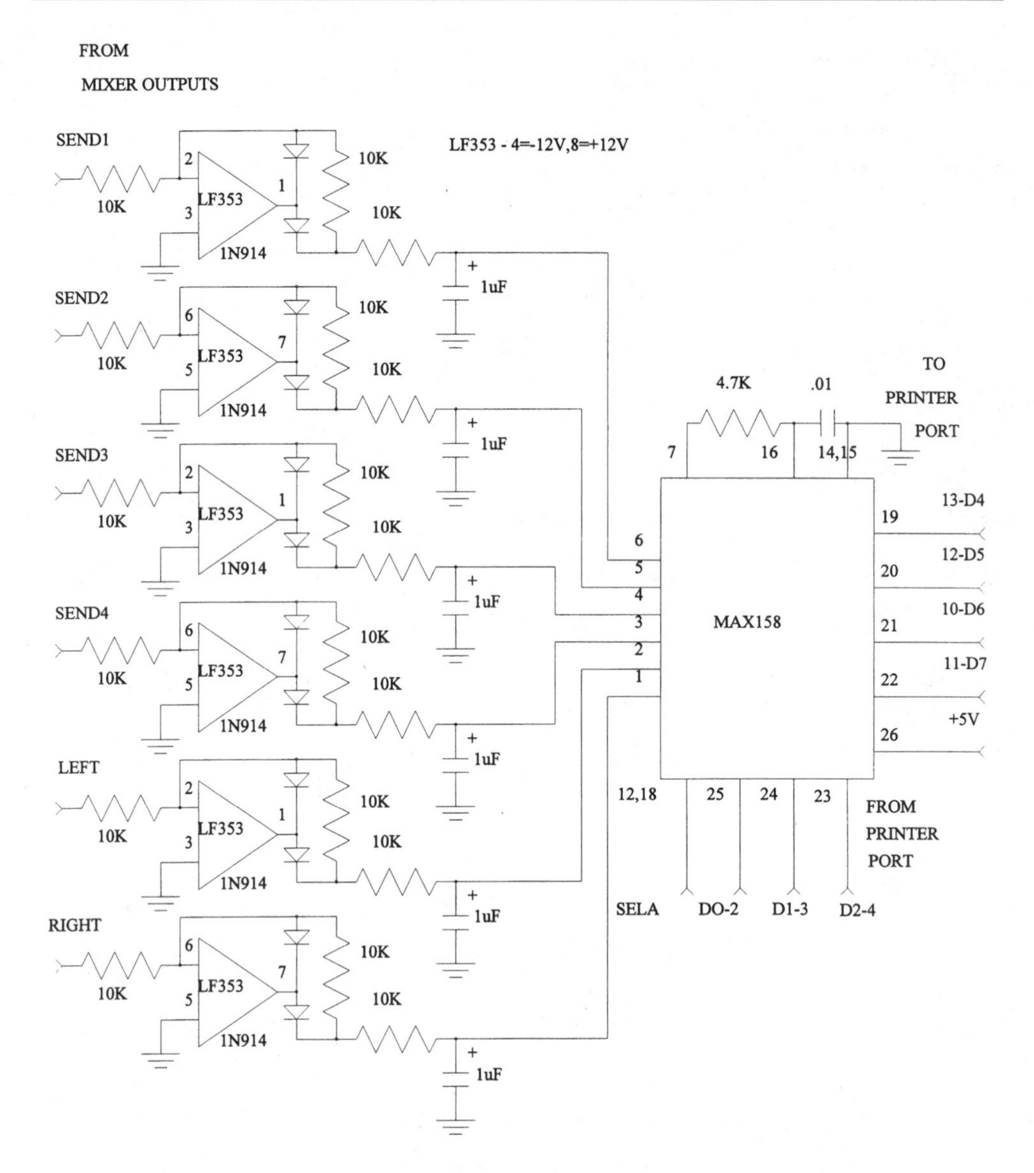

Figure 2-29: Printer port audio mixer analog-to-digital converter

Get ready. The software for this project is quite extensive. There are some fancy features, such as the ability to display round knobs and make them rotate under software control. A ">" sign shows which control is selected and currently being

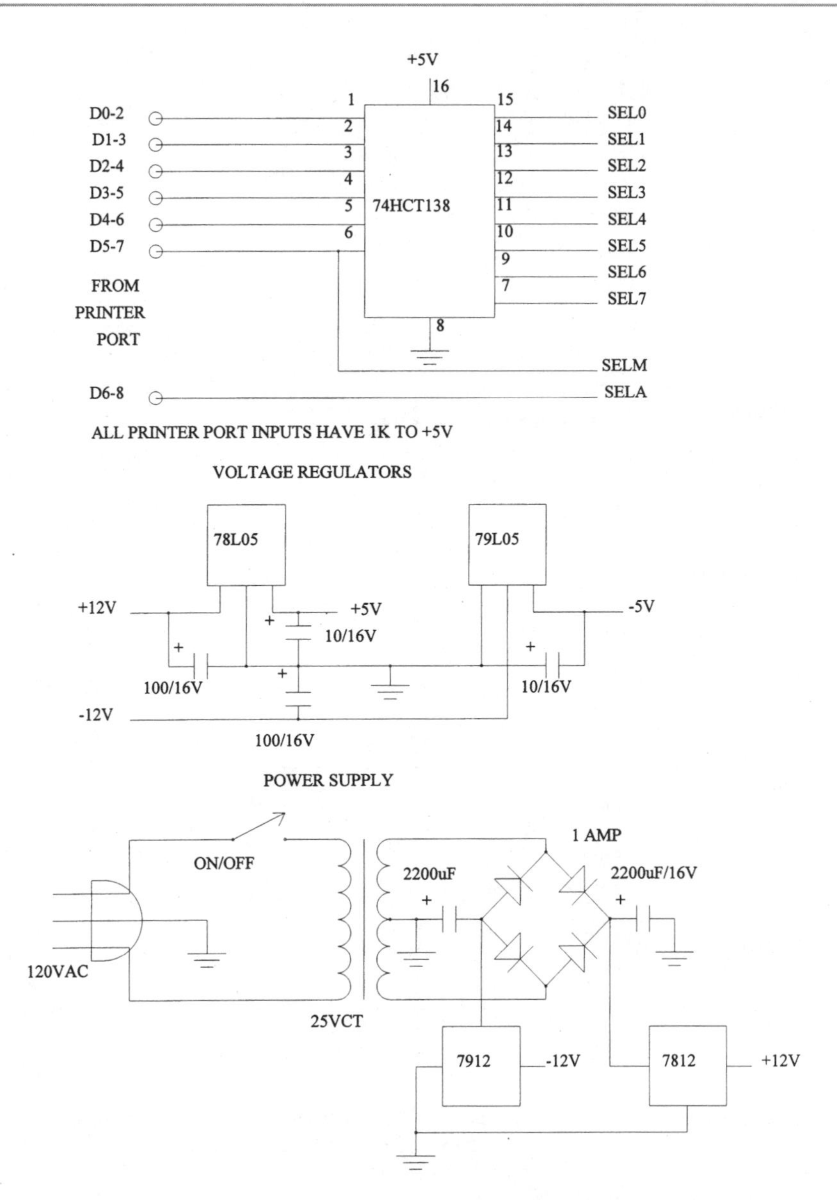

Figure 2-30: Printer port audio mixer select and power circuits

adjusted. The function keys F1 to F9 select the channel to be adjusted. The F9 key selects the master channel. The up and down arrow keys select which control to be adjusted. The right and left arrow keys actually turn the knob.

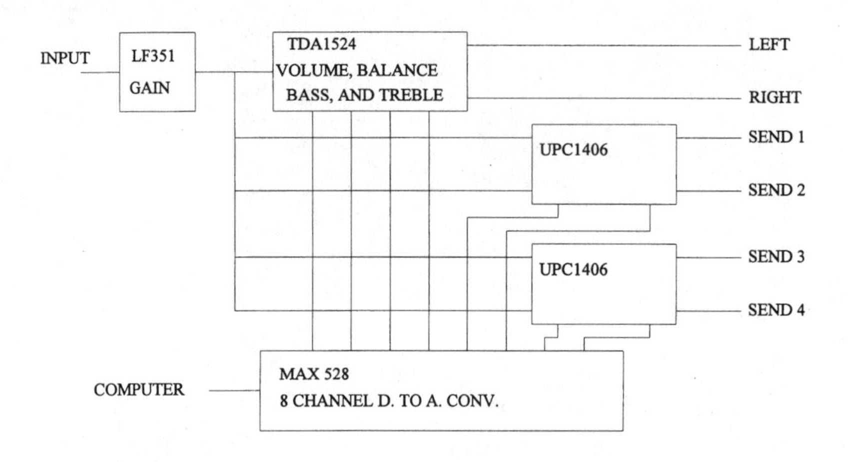

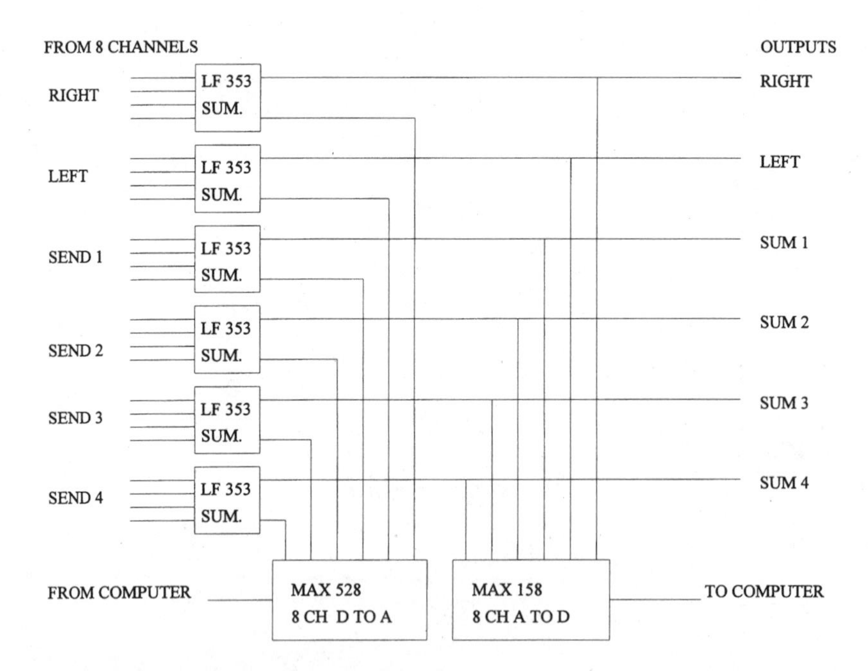

Figure 2-31: Printer port audio mixer block diagrams

There are two versions of the software: one for CGA and one for VGA.

```
' PRINTER PORT AUDIO MIXER  6/23/95 BY BOB DAVIS
' LOW RESOLUTION VERSION FOR 640 X 200 CGA
' DOUT=DATA OUT, COUT=CONTROL OUT, CIN=CONTROL IN
CLS
LOCATE 8, 20: PRINT "COMPUTER CONTROLLED AUDIO MIXER"
LOCATE 11, 20: PRINT "COPYRIGHT 1995 BY ROBERT J DAVIS"
LOCATE 14, 20: INPUT "USE PRINTER PORT NUMBER: ", LPT
DOUT = &H378: COUT = &H37A: CIN = &H379
IF LPT = 1 THEN DOUT = &H3BC: COUT = &H3BE: CIN = &H3BD
IF LPT = 3 THEN DOUT = &H278: COUT = &H27A: CIN = &H279
CLS : SCREEN 2: KNOB = 1: CHANNEL = 1
DRAW "BM 0,0" + "R 639" + "D 189" + "L 639" + "U 189"
DRAW "BM 0,0" + "R 400" + "D 189"
LOCATE 3, 44: PRINT "TREB": LOCATE 5, 44: PRINT "BASS"
LOCATE 7, 44: PRINT "MON.": LOCATE 9, 44: PRINT "EFF."
LOCATE 11, 44: PRINT "SEND": LOCATE 13, 44: PRINT "QUE"
LOCATE 15, 44: PRINT "BAL.": LOCATE 23, 44: PRINT "CHAN"
LOCATE 23, 5: PRINT "1     2     3     4     5     6     7     8"
LOCATE 15, 52: PRINT "MON. EFF. SEND  QUE  LT  RT"
LOCATE 25, 1: PRINT "F1 - F9 = CHANNEL, ARROW KEYS = CONTROLS, F=FILE,
  Q=QUIT";
T1 = 128: T2 = 128: T3 = 128: T4 = 128: T5 = 128: T6 = 128: T7 = 128: T8 =
  128
B1 = 128: B2 = 128: B3 = 128: B4 = 128: B5 = 128: B6 = 128: B7 = 128: B8 =
  128
C1 = 128: C2 = 128: C3 = 128: C4 = 128: C5 = 128: C6 = 128: C7 = 128: C8 =
  128
V1 = 0: V2 = 0: V3 = 0: V4 = 0: V5 = 0: V6 = 0: V7 = 0: V8 = 0
M1 = 0: M2 = 0: M3 = 0: M4 = 0: M5 = 0: M6 = 0: M7 = 0: M8 = 0
E1 = 0: E2 = 0: E3 = 0: E4 = 0: E5 = 0: E6 = 0: E7 = 0: E8 = 0
S1 = 0: S2 = 0: S3 = 0: S4 = 0: S5 = 0: S6 = 0: S7 = 0: S8 = 0
Q1 = 0: Q2 = 0: Q3 = 0: Q4 = 0: Q5 = 0: Q6 = 0: Q7 = 0: Q8 = 0
LT = 128: RT = 128: MO = 128: EF = 128: SE = 128: QU = 128
start:
FOR B = 35 TO 630 STEP 40     ' DRAW SLIDE CONTROLS
IF B < 340 OR B > 400 THEN
   PSET (B, 126): LINE -STEP(0, 48)
END IF
IF B = 395 THEN B = 380
IF B > 380 THEN
   'IF B <> 500 AND B <> 540 THEN
      FOR A = 17 TO 110 STEP 6
         CIRCLE (B, A), 5
      NEXT A
   'END IF
END IF
NEXT B
PSET (27, 167 - (V1 / 6)): DRAW "C0" + "M+16,+0" + "M+0,+6" + "M-16,+0" +
  "M+0,-6"
```

```
PSET (27, 169 - (V1 / 6)): DRAW "C0" + "M+16,+0" + "M+0,+6" + "M-16,+0" +
   "M+0,-6"
PSET (27, 168 - (V1 / 6)): DRAW "C1" + "M+16,+0" + "M+0,+6" + "M-16,+0" +
   "M+0,-6"
PSET (67, 167 - (V2 / 6)): DRAW "C0" + "M+16,+0" + "M+0,+6" + "M-16,+0" +
   "M+0,-6"
PSET (67, 169 - (V2 / 6)): DRAW "C0" + "M+16,+0" + "M+0,+6" + "M-16,+0" +
   "M+0,-6"
PSET (67, 168 - (V2 / 6)): DRAW "C1" + "M+16,+0" + "M+0,+6" + "M-16,+0" +
   "M+0,-6"
PSET (107, 167 - (V3 / 6)): DRAW "C0" + "M+16,+0" + "M+0,+6" + "M-16,+0" +
   "M+0,-6"
PSET (107, 169 - (V3 / 6)): DRAW "C0" + "M+16,+0" + "M+0,+6" + "M-16,+0" +
   "M+0,-6"
PSET (107, 168 - (V3 / 6)): DRAW "C1" + "M+16,+0" + "M+0,+6" + "M-16,+0" +
   "M+0,-6"
PSET (147, 167 - (V4 / 6)): DRAW "C0" + "M+16,+0" + "M+0,+6" + "M-16,+0" +
   "M+0,-6"
PSET (147, 169 - (V4 / 6)): DRAW "C0" + "M+16,+0" + "M+0,+6" + "M-16,+0" +
   "M+0,-6"
PSET (147, 168 - (V4 / 6)): DRAW "C1" + "M+16,+0" + "M+0,+6" + "M-16,+0" +
   "M+0,-6"
PSET (187, 167 - (V5 / 6)): DRAW "C0" + "M+16,+0" + "M+0,+6" + "M-16,+0" +
   "M+0,-6"
PSET (187, 169 - (V5 / 6)): DRAW "C0" + "M+16,+0" + "M+0,+6" + "M-16,+0" +
   "M+0,-6"
PSET (187, 168 - (V5 / 6)): DRAW "C1" + "M+16,+0" + "M+0,+6" + "M-16,+0" +
   "M+0,-6"
PSET (227, 167 - (V6 / 6)): DRAW "C0" + "M+16,+0" + "M+0,+6" + "M-16,+0" +
   "M+0,-6"
PSET (227, 169 - (V6 / 6)): DRAW "C0" + "M+16,+0" + "M+0,+6" + "M-16,+0" +
   "M+0,-6"
PSET (227, 168 - (V6 / 6)): DRAW "C1" + "M+16,+0" + "M+0,+6" + "M-16,+0" +
   "M+0,-6"
PSET (267, 167 - (V7 / 6)): DRAW "C0" + "M+16,+0" + "M+0,+6" + "M-16,+0" +
   "M+0,-6"
PSET (267, 169 - (V7 / 6)): DRAW "C0" + "M+16,+0" + "M+0,+6" + "M-16,+0" +
   "M+0,-6"
PSET (267, 168 - (V7 / 6)): DRAW "C1" + "M+16,+0" + "M+0,+6" + "M-16,+0" +
   "M+0,-6"
PSET (307, 167 - (V8 / 6)): DRAW "C0" + "M+16,+0" + "M+0,+6" + "M-16,+0" +
   "M+0,-6"
PSET (307, 169 - (V8 / 6)): DRAW "C0" + "M+16,+0" + "M+0,+6" + "M-16,+0" +
   "M+0,-6"
PSET (307, 168 - (V8 / 6)): DRAW "C1" + "M+16,+0" + "M+0,+6" + "M-16,+0" +
   "M+0,-6"
PSET (412, 167 - (MO / 6)): DRAW "C0" + "M+16,+0" + "M+0,+6" + "M-16,+0" +
   "M+0,-6"
PSET (412, 169 - (MO / 6)): DRAW "C0" + "M+16,+0" + "M+0,+6" + "M-16,+0" +
   "M+0,-6"
PSET (412, 168 - (MO / 6)): DRAW "C1" + "M+16,+0" + "M+0,+6" + "M-16,+0" +
   "M+0,-6"
```

```
PSET (452, 167 - (EF / 6)): DRAW "C0" + "M+16,+0" + "M+0,+6" + "M-16,+0" +
   "M+0,-6"
PSET (452, 169 - (EF / 6)): DRAW "C0" + "M+16,+0" + "M+0,+6" + "M-16,+0" +
   "M+0,-6"
PSET (452, 168 - (EF / 6)): DRAW "C1" + "M+16,+0" + "M+0,+6" + "M-16,+0" +
   "M+0,-6"
PSET (492, 167 - (SE / 6)): DRAW "C0" + "M+16,+0" + "M+0,+6" + "M-16,+0" +
   "M+0,-6"
PSET (492, 169 - (SE / 6)): DRAW "C0" + "M+16,+0" + "M+0,+6" + "M-16,+0" +
   "M+0,-6"
PSET (492, 168 - (SE / 6)): DRAW "C1" + "M+16,+0" + "M+0,+6" + "M-16,+0" +
   "M+0,-6"
PSET (532, 167 - (QU / 6)): DRAW "C0" + "M+16,+0" + "M+0,+6" + "M-16,+0" +
   "M+0,-6"
PSET (532, 169 - (QU / 6)): DRAW "C0" + "M+16,+0" + "M+0,+6" + "M-16,+0" +
   "M+0,-6"
PSET (532, 168 - (QU / 6)): DRAW "C1" + "M+16,+0" + "M+0,+6" + "M-16,+0" +
   "M+0,-6"
PSET (572, 167 - (LT / 6)): DRAW "C0" + "M+16,+0" + "M+0,+6" + "M-16,+0" +
   "M+0,-6"
PSET (572, 169 - (LT / 6)): DRAW "C0" + "M+16,+0" + "M+0,+6" + "M-16,+0" +
   "M+0,-6"
PSET (572, 168 - (LT / 6)): DRAW "C1" + "M+16,+0" + "M+0,+6" + "M-16,+0" +
   "M+0,-6"
PSET (612, 167 - (RT / 6)): DRAW "C0" + "M+16,+0" + "M+0,+6" + "M-16,+0" +
   "M+0,-6"
PSET (612, 169 - (RT / 6)): DRAW "C0" + "M+16,+0" + "M+0,+6" + "M-16,+0" +
   "M+0,-6"
PSET (612, 168 - (RT / 6)): DRAW "C1" + "M+16,+0" + "M+0,+6" + "M-16,+0" +
   "M+0,-6"
FOR A = 35 TO 335 STEP 40     ' DRAW KNOB CONTROLS
FOR B = 19 TO 115 STEP 16
    IF B = 19 AND A = 35 THEN CONTROL = T1
    IF B = 19 AND A = 75 THEN CONTROL = T2
    IF B = 19 AND A = 115 THEN CONTROL = T3
    IF B = 19 AND A = 155 THEN CONTROL = T4
    IF B = 19 AND A = 195 THEN CONTROL = T5
    IF B = 19 AND A = 235 THEN CONTROL = T6
    IF B = 19 AND A = 275 THEN CONTROL = T7
    IF B = 19 AND A = 315 THEN CONTROL = T8
    IF B = 35 AND A = 35 THEN CONTROL = B1
    IF B = 35 AND A = 75 THEN CONTROL = B2
    IF B = 35 AND A = 115 THEN CONTROL = B3
    IF B = 35 AND A = 155 THEN CONTROL = B4
    IF B = 35 AND A = 195 THEN CONTROL = B5
    IF B = 35 AND A = 235 THEN CONTROL = B6
    IF B = 35 AND A = 275 THEN CONTROL = B7
    IF B = 35 AND A = 315 THEN CONTROL = B8
    IF B = 51 AND A = 35 THEN CONTROL = M1
    IF B = 51 AND A = 75 THEN CONTROL = M2
    IF B = 51 AND A = 115 THEN CONTROL = M3
    IF B = 51 AND A = 155 THEN CONTROL = M4
```

```
IF B = 51 AND A = 195 THEN CONTROL = M5
IF B = 51 AND A = 235 THEN CONTROL = M6
IF B = 51 AND A = 275 THEN CONTROL = M7
IF B = 51 AND A = 315 THEN CONTROL = M8
IF B = 67 AND A = 35 THEN CONTROL = E1
IF B = 67 AND A = 75 THEN CONTROL = E2
IF B = 67 AND A = 115 THEN CONTROL = E3
IF B = 67 AND A = 155 THEN CONTROL = E4
IF B = 67 AND A = 195 THEN CONTROL = E5
IF B = 67 AND A = 235 THEN CONTROL = E6
IF B = 67 AND A = 275 THEN CONTROL = E7
IF B = 67 AND A = 315 THEN CONTROL = E8
IF B = 83 AND A = 35 THEN CONTROL = S1
IF B = 83 AND A = 75 THEN CONTROL = S2
IF B = 83 AND A = 115 THEN CONTROL = S3
IF B = 83 AND A = 155 THEN CONTROL = S4
IF B = 83 AND A = 195 THEN CONTROL = S5
IF B = 83 AND A = 235 THEN CONTROL = S6
IF B = 83 AND A = 275 THEN CONTROL = S7
IF B = 83 AND A = 315 THEN CONTROL = S8
IF B = 99 AND A = 35 THEN CONTROL = Q1
IF B = 99 AND A = 75 THEN CONTROL = Q2
IF B = 99 AND A = 115 THEN CONTROL = Q3
IF B = 99 AND A = 155 THEN CONTROL = Q4
IF B = 99 AND A = 195 THEN CONTROL = Q5
IF B = 99 AND A = 235 THEN CONTROL = Q6
IF B = 99 AND A = 275 THEN CONTROL = Q7
IF B = 99 AND A = 315 THEN CONTROL = Q8
IF B = 115 AND A = 35 THEN CONTROL = C1
IF B = 115 AND A = 75 THEN CONTROL = C2
IF B = 115 AND A = 115 THEN CONTROL = C3
IF B = 115 AND A = 155 THEN CONTROL = C4
IF B = 115 AND A = 195 THEN CONTROL = C5
IF B = 115 AND A = 235 THEN CONTROL = C6
IF B = 115 AND A = 275 THEN CONTROL = C7
IF B = 115 AND A = 315 THEN CONTROL = C8
CIRCLE (A, B), 12
SELECT CASE CONTROL
CASE IS < 32
   PSET (A, B): LINE -STEP(6, 4), 0
   PSET (A, B): LINE -STEP(-10, 0), 0
   PSET (A, B): LINE -STEP(-6, 4)
CASE IS < 64
   PSET (A, B): LINE -STEP(-6, 4), 0
   PSET (A, B): LINE -STEP(-6, -4), 0
   PSET (A, B): LINE -STEP(-10, 0)
CASE IS < 96
   PSET (A, B): LINE -STEP(-10, 0), 0
   PSET (A, B): LINE -STEP(0, -4), 0
   PSET (A, B): LINE -STEP(-6, -4)
CASE IS < 160
   PSET (A, B): LINE -STEP(-6, -4), 0
   PSET (A, B): LINE -STEP(6, -4), 0
```

```
      PSET (A, B): LINE -STEP(0, -4)
   CASE IS < 192
      PSET (A, B): LINE -STEP(0, -4), 0
      PSET (A, B): LINE -STEP(10, 0), 0
      PSET (A, B): LINE -STEP(6, -4)
   CASE IS < 224
      PSET (A, B): LINE -STEP(6, -4), 0
      PSET (A, B): LINE -STEP(6, 4), 0
      PSET (A, B): LINE -STEP(10, 0)
   CASE IS < 256
      PSET (A, B): LINE -STEP(10, 0), 0
      PSET (A, B): LINE -STEP(-6, 4), 0
      PSET (A, B): LINE -STEP(6, 4)
   END SELECT
NEXT B
NEXT A
LOCATE KNOB * 2 + 1, (CHANNEL * 5 - 2): PRINT " "
KEY$ = INKEY$
IF KEY$ = "q" OR KEY$ = "Q" THEN END
IF LEN(KEY$) > 1 THEN                    ' FUNCTION KEY INTERPRETER
IF RIGHT$(KEY$, 1) = ";" THEN CHANNEL = 1
IF RIGHT$(KEY$, 1) = "<" THEN CHANNEL = 2
IF RIGHT$(KEY$, 1) = "=" THEN CHANNEL = 3
IF RIGHT$(KEY$, 1) = ">" THEN CHANNEL = 4
IF RIGHT$(KEY$, 1) = "?" THEN CHANNEL = 5
IF RIGHT$(KEY$, 1) = "@" THEN CHANNEL = 6
IF RIGHT$(KEY$, 1) = "A" THEN CHANNEL = 7
IF RIGHT$(KEY$, 1) = "B" THEN CHANNEL = 8
IF RIGHT$(KEY$, 1) = "C" THEN CHANNEL = 9
END IF
IF KEY$ = "8" THEN
KNOB = KNOB - 1
IF KNOB = 0 THEN KNOB = 8
END IF
IF KEY$ = "2" THEN
KNOB = KNOB + 1
IF KNOB = 9 THEN KNOB = 1
END IF                                    ' DRAW SELECT POINTER ARROW
LOCATE KNOB * 2 + 1, (CHANNEL * 5 - 2): PRINT ">"
IF CHANNEL = 1 AND KNOB = 8 THEN CONTROL = V1
IF CHANNEL = 1 AND KNOB = 7 THEN CONTROL = C1
IF CHANNEL = 1 AND KNOB = 6 THEN CONTROL = Q1
IF CHANNEL = 1 AND KNOB = 5 THEN CONTROL = S1
IF CHANNEL = 1 AND KNOB = 4 THEN CONTROL = E1
IF CHANNEL = 1 AND KNOB = 3 THEN CONTROL = M1
IF CHANNEL = 1 AND KNOB = 2 THEN CONTROL = B1
IF CHANNEL = 1 AND KNOB = 1 THEN CONTROL = T1
IF CHANNEL = 2 AND KNOB = 8 THEN CONTROL = V2
IF CHANNEL = 2 AND KNOB = 7 THEN CONTROL = C2
IF CHANNEL = 2 AND KNOB = 6 THEN CONTROL = Q2
IF CHANNEL = 2 AND KNOB = 5 THEN CONTROL = S2
IF CHANNEL = 2 AND KNOB = 4 THEN CONTROL = E2
IF CHANNEL = 2 AND KNOB = 3 THEN CONTROL = M2
```

```
IF CHANNEL = 2 AND KNOB = 2 THEN CONTROL = B2
IF CHANNEL = 2 AND KNOB = 1 THEN CONTROL = T2
IF CHANNEL = 3 AND KNOB = 8 THEN CONTROL = V3
IF CHANNEL = 3 AND KNOB = 7 THEN CONTROL = C3
IF CHANNEL = 3 AND KNOB = 6 THEN CONTROL = Q3
IF CHANNEL = 3 AND KNOB = 5 THEN CONTROL = S3
IF CHANNEL = 3 AND KNOB = 4 THEN CONTROL = E3
IF CHANNEL = 3 AND KNOB = 3 THEN CONTROL = M3
IF CHANNEL = 3 AND KNOB = 2 THEN CONTROL = B3
IF CHANNEL = 3 AND KNOB = 1 THEN CONTROL = T3
IF CHANNEL = 4 AND KNOB = 8 THEN CONTROL = V4
IF CHANNEL = 4 AND KNOB = 7 THEN CONTROL = C4
IF CHANNEL = 4 AND KNOB = 6 THEN CONTROL = Q4
IF CHANNEL = 4 AND KNOB = 5 THEN CONTROL = S4
IF CHANNEL = 4 AND KNOB = 4 THEN CONTROL = E4
IF CHANNEL = 4 AND KNOB = 3 THEN CONTROL = M4
IF CHANNEL = 4 AND KNOB = 2 THEN CONTROL = B4
IF CHANNEL = 4 AND KNOB = 1 THEN CONTROL = T4
IF CHANNEL = 5 AND KNOB = 8 THEN CONTROL = V5
IF CHANNEL = 5 AND KNOB = 7 THEN CONTROL = C5
IF CHANNEL = 5 AND KNOB = 6 THEN CONTROL = Q5
IF CHANNEL = 5 AND KNOB = 5 THEN CONTROL = S5
IF CHANNEL = 5 AND KNOB = 4 THEN CONTROL = E5
IF CHANNEL = 5 AND KNOB = 3 THEN CONTROL = M5
IF CHANNEL = 5 AND KNOB = 2 THEN CONTROL = B5
IF CHANNEL = 5 AND KNOB = 1 THEN CONTROL = T5
IF CHANNEL = 6 AND KNOB = 8 THEN CONTROL = V6
IF CHANNEL = 6 AND KNOB = 7 THEN CONTROL = C6
IF CHANNEL = 6 AND KNOB = 6 THEN CONTROL = Q6
IF CHANNEL = 6 AND KNOB = 5 THEN CONTROL = S6
IF CHANNEL = 6 AND KNOB = 4 THEN CONTROL = E6
IF CHANNEL = 6 AND KNOB = 3 THEN CONTROL = M6
IF CHANNEL = 6 AND KNOB = 2 THEN CONTROL = B6
IF CHANNEL = 6 AND KNOB = 1 THEN CONTROL = T6
IF CHANNEL = 7 AND KNOB = 8 THEN CONTROL = V7
IF CHANNEL = 7 AND KNOB = 7 THEN CONTROL = C7
IF CHANNEL = 7 AND KNOB = 6 THEN CONTROL = Q7
IF CHANNEL = 7 AND KNOB = 5 THEN CONTROL = S7
IF CHANNEL = 7 AND KNOB = 4 THEN CONTROL = E7
IF CHANNEL = 7 AND KNOB = 3 THEN CONTROL = M7
IF CHANNEL = 7 AND KNOB = 2 THEN CONTROL = B7
IF CHANNEL = 7 AND KNOB = 1 THEN CONTROL = T7
IF CHANNEL = 8 AND KNOB = 8 THEN CONTROL = V8
IF CHANNEL = 8 AND KNOB = 7 THEN CONTROL = C8
IF CHANNEL = 8 AND KNOB = 6 THEN CONTROL = Q8
IF CHANNEL = 8 AND KNOB = 5 THEN CONTROL = S8
IF CHANNEL = 8 AND KNOB = 4 THEN CONTROL = E8
IF CHANNEL = 8 AND KNOB = 3 THEN CONTROL = M8
IF CHANNEL = 8 AND KNOB = 2 THEN CONTROL = B8
IF CHANNEL = 8 AND KNOB = 1 THEN CONTROL = T8
IF CHANNEL = 9 AND KNOB = 8 THEN CONTROL = RT
IF CHANNEL = 9 AND KNOB = 7 THEN CONTROL = LT
```

```
IF CHANNEL = 9 AND KNOB = 6 THEN CONTROL = QU
IF CHANNEL = 9 AND KNOB = 5 THEN CONTROL = SE
IF CHANNEL = 9 AND KNOB = 4 THEN CONTROL = EF
IF CHANNEL = 9 AND KNOB = 3 THEN CONTROL = MO
IF CHANNEL = 9 AND KNOB = 2 THEN CONTROL = RT
IF CHANNEL = 9 AND KNOB = 1 THEN CONTROL = LT
IF KEY$ = "4" THEN
   CONTROL = CONTROL - 4
   IF CONTROL < 0 THEN CONTROL = 0
END IF
IF KEY$ = "6" THEN
   CONTROL = CONTROL + 4
   IF CONTROL > 252 THEN CONTROL = 252
END IF                                   ' DRAW SELECT POINTER ARROW
IF CHANNEL = 1 AND KNOB = 8 THEN V1 = CONTROL
IF CHANNEL = 1 AND KNOB = 7 THEN C1 = CONTROL
IF CHANNEL = 1 AND KNOB = 6 THEN Q1 = CONTROL
IF CHANNEL = 1 AND KNOB = 5 THEN S1 = CONTROL
IF CHANNEL = 1 AND KNOB = 4 THEN E1 = CONTROL
IF CHANNEL = 1 AND KNOB = 3 THEN M1 = CONTROL
IF CHANNEL = 1 AND KNOB = 2 THEN B1 = CONTROL
IF CHANNEL = 1 AND KNOB = 1 THEN T1 = CONTROL
IF CHANNEL = 2 AND KNOB = 8 THEN V2 = CONTROL
IF CHANNEL = 2 AND KNOB = 7 THEN C2 = CONTROL
IF CHANNEL = 2 AND KNOB = 6 THEN Q2 = CONTROL
IF CHANNEL = 2 AND KNOB = 5 THEN S2 = CONTROL
IF CHANNEL = 2 AND KNOB = 4 THEN E2 = CONTROL
IF CHANNEL = 2 AND KNOB = 3 THEN M2 = CONTROL
IF CHANNEL = 2 AND KNOB = 2 THEN B2 = CONTROL
IF CHANNEL = 2 AND KNOB = 1 THEN T2 = CONTROL
IF CHANNEL = 3 AND KNOB = 8 THEN V3 = CONTROL
IF CHANNEL = 3 AND KNOB = 7 THEN C3 = CONTROL
IF CHANNEL = 3 AND KNOB = 6 THEN Q3 = CONTROL
IF CHANNEL = 3 AND KNOB = 5 THEN S3 = CONTROL
IF CHANNEL = 3 AND KNOB = 4 THEN E3 = CONTROL
IF CHANNEL = 3 AND KNOB = 3 THEN M3 = CONTROL
IF CHANNEL = 3 AND KNOB = 2 THEN B3 = CONTROL
IF CHANNEL = 3 AND KNOB = 1 THEN T3 = CONTROL
IF CHANNEL = 4 AND KNOB = 8 THEN V4 = CONTROL
IF CHANNEL = 4 AND KNOB = 7 THEN C4 = CONTROL
IF CHANNEL = 4 AND KNOB = 6 THEN Q4 = CONTROL
IF CHANNEL = 4 AND KNOB = 5 THEN S4 = CONTROL
IF CHANNEL = 4 AND KNOB = 4 THEN E4 = CONTROL
IF CHANNEL = 4 AND KNOB = 3 THEN M4 = CONTROL
IF CHANNEL = 4 AND KNOB = 2 THEN B4 = CONTROL
IF CHANNEL = 4 AND KNOB = 1 THEN T4 = CONTROL
IF CHANNEL = 5 AND KNOB = 8 THEN V5 = CONTROL
IF CHANNEL = 5 AND KNOB = 7 THEN C5 = CONTROL
IF CHANNEL = 5 AND KNOB = 6 THEN Q5 = CONTROL
IF CHANNEL = 5 AND KNOB = 5 THEN S5 = CONTROL
IF CHANNEL = 5 AND KNOB = 4 THEN E5 = CONTROL
IF CHANNEL = 5 AND KNOB = 3 THEN M5 = CONTROL
```

```
IF CHANNEL = 5 AND KNOB = 2 THEN B5 = CONTROL
IF CHANNEL = 5 AND KNOB = 1 THEN T5 = CONTROL
IF CHANNEL = 6 AND KNOB = 8 THEN V6 = CONTROL
IF CHANNEL = 6 AND KNOB = 7 THEN C6 = CONTROL
IF CHANNEL = 6 AND KNOB = 6 THEN Q6 = CONTROL
IF CHANNEL = 6 AND KNOB = 5 THEN S6 = CONTROL
IF CHANNEL = 6 AND KNOB = 4 THEN E6 = CONTROL
IF CHANNEL = 6 AND KNOB = 3 THEN M6 = CONTROL
IF CHANNEL = 6 AND KNOB = 2 THEN B6 = CONTROL
IF CHANNEL = 6 AND KNOB = 1 THEN T6 = CONTROL
IF CHANNEL = 7 AND KNOB = 8 THEN V7 = CONTROL
IF CHANNEL = 7 AND KNOB = 7 THEN C7 = CONTROL
IF CHANNEL = 7 AND KNOB = 6 THEN Q7 = CONTROL
IF CHANNEL = 7 AND KNOB = 5 THEN S7 = CONTROL
IF CHANNEL = 7 AND KNOB = 4 THEN E7 = CONTROL
IF CHANNEL = 7 AND KNOB = 3 THEN M7 = CONTROL
IF CHANNEL = 7 AND KNOB = 2 THEN B7 = CONTROL
IF CHANNEL = 7 AND KNOB = 1 THEN T7 = CONTROL
IF CHANNEL = 8 AND KNOB = 8 THEN V8 = CONTROL
IF CHANNEL = 8 AND KNOB = 7 THEN C8 = CONTROL
IF CHANNEL = 8 AND KNOB = 6 THEN Q8 = CONTROL
IF CHANNEL = 8 AND KNOB = 5 THEN S8 = CONTROL
IF CHANNEL = 8 AND KNOB = 4 THEN E8 = CONTROL
IF CHANNEL = 8 AND KNOB = 3 THEN M8 = CONTROL
IF CHANNEL = 8 AND KNOB = 2 THEN B8 = CONTROL
IF CHANNEL = 8 AND KNOB = 1 THEN T8 = CONTROL
IF CHANNEL = 9 AND KNOB = 8 THEN RT = CONTROL
IF CHANNEL = 9 AND KNOB = 7 THEN LT = CONTROL
IF CHANNEL = 9 AND KNOB = 6 THEN QU = CONTROL
IF CHANNEL = 9 AND KNOB = 5 THEN SE = CONTROL
IF CHANNEL = 9 AND KNOB = 4 THEN EF = CONTROL
IF CHANNEL = 9 AND KNOB = 3 THEN MO = CONTROL
IF CHANNEL = 9 AND KNOB = 2 THEN RT = CONTROL
IF CHANNEL = 9 AND KNOB = 1 THEN LT = CONTROL
LOCATE 25, 62: PRINT "SETTING=";
LOCATE 25, 70: PRINT CONTROL;
'********** SEND THE INFORMATION TO THE MIXER **********
'CONT OUT D0 (1) = CLOCK, D1 (2) = DATA, D2 (4\) MASTER, D3 (8) A:D ENABLE
IF CHANNEL = 1 THEN OUT DOUT, &HFE
IF CHANNEL = 2 THEN OUT DOUT, &HFD
IF CHANNEL = 3 THEN OUT DOUT, &HFB
IF CHANNEL = 4 THEN OUT DOUT, &HF7
IF CHANNEL = 5 THEN OUT DOUT, &HEF
IF CHANNEL = 6 THEN OUT DOUT, &HDF
IF CHANNEL = 7 THEN OUT DOUT, &HBF
IF CHANNEL = 8 THEN OUT DOUT, &H7F
IF CHANNEL = 9 THEN BITADD = 0 ELSE BITADD = 4
NEWKNOB = KNOB - 2: IF NEWKNOB < 1 THEN NEWKNOB = NEWKNOB + 8
FOR A = 1 TO 8
IF A = NEWKNOB THEN      ' SELECTS CONTROL TO CHANGE REVERSE (1=8)
   OUT COUT, 1 + BITADD: OUT COUT, 0 + BITADD
ELSE
```

```
    OUT COUT, 3 + BITADD: OUT COUT, 2 + BITADD
END IF
NEXT A
FOR A = 1 TO 8              ' SERIALLY SEND SELECTED CONTROL
BITOUT = 0
IF A = 1 AND (CONTROL AND 128) = 128 THEN BITOUT = 1
IF A = 2 AND (CONTROL AND 64) = 64 THEN BITOUT = 1
IF A = 3 AND (CONTROL AND 32) = 32 THEN BITOUT = 1
IF A = 4 AND (CONTROL AND 16) = 16 THEN BITOUT = 1
IF A = 5 AND (CONTROL AND 8) = 8 THEN BITOUT = 1
IF A = 6 AND (CONTROL AND 4) = 4 THEN BITOUT = 1
IF A = 7 AND (CONTROL AND 2) = 2 THEN BITOUT = 1
IF A = 8 AND (CONTROL AND 1) = 1 THEN BITOUT = 1
IF BITOUT = 1 THEN OUT COUT, 1 + BITADD: OUT COUT, 0 + BITADD
IF BITOUT = 0 THEN OUT COUT, 3 + BITADD: OUT COUT, 2 + BITADD
NEXT A
OUT DOUT, &HFF                 ' DESELECT ALL D:A CONVERTERS
OUT COUT, 4                    ' DESELECT MASTER D:A CONVERTER
'********** DISPLAY RESULTS IN THE VU METERS **********
METER = 3
FOR B = 420 TO 620 STEP 40    ' DRAW VU METERS
IF B = 500 THEN B = 580
OUT DOUT, METER
OUT COUT, 15                ' SELECT A:D CONVERTER DATA & CLOCK = "0"
VUMETER = (INP(CIN) XOR &H80) \ 16' INVERT BIT 7 & SHIFT DATA
AV = 15
FOR A = 17 TO 110 STEP 6
   IF VUMETER > AV THEN PAINT (B, A), 2 ELSE PAINT (B, A), 0
   AV = AV - 1
NEXT A
METER = METER - 1
OUT COUT, 7                    ' DE-SELECT A:D CONVERTER
NEXT B
OUT DOUT, &HFF                 ' DESELECT ALL D:A CONVERTERS
GOTO start
```

This is the high-resolution version. It is newer and the amount of code has been condensed.

```
CLS : LOCATE 8, 20: PRINT "COMPUTER CONTROLLED AUDIO MIXER"
LOCATE 11, 20: PRINT "COPYRIGHT 1995 BY ROBERT J DAVIS"
LOCATE 14, 20: INPUT "SELECT PRINTER PORT NUMBER: ", LPT
DOUT = &H378: IF LPT = 1 THEN DOUT = &H3BC
IF LPT = 3 THEN DOUT = &H278
CIN = DOUT + 1: COUT = DOUT + 2
CLS : SCREEN 12: KNOB = 1: CHAN = 1: DIM C(8, 9)
DRAW "BM 0,0" + "R 639" + "D 440" + "L 639" + "U 440" + "R 360" + "D 440"
LINE (17, 333)-(623, 333): LINE (17, 349)-(623, 349)
LINE (17, 365)-(623, 365): LINE (17, 380)-(623, 380)
LINE (17, 397)-(623, 397): LINE (17, 415)-(623, 415)
LOCATE 3, 44: PRINT "TREB": LOCATE 6, 44: PRINT "BASS"
```

```
LOCATE 8, 44: PRINT "SND1": LOCATE 11, 44: PRINT "SND2"
LOCATE 13, 44: PRINT "SND3": LOCATE 16, 44: PRINT "SND4"
LOCATE 18, 44: PRINT "BAL": LOCATE 20, 44: PRINT "VOL"
LOCATE 21, 44: PRINT "+20 ": LOCATE 22, 44: PRINT "  0 "
LOCATE 23, 44: PRINT "-20 ": LOCATE 24, 44: PRINT "-40 "
LOCATE 25, 44: PRINT "-60 ": LOCATE 26, 44: PRINT "-80 "
LOCATE 20, 6: PRINT "1    2    3    4    5    6    7    8"
LOCATE 20, 49: PRINT "SND1 SND2 SND3 SND4 LEFT RIGHT"
START: LOCATE 29, 1: PRINT "ARROWS SELECT, +,- ADJUST, S=SAVE,";
LOCATE 29, 36: PRINT "R=RECALL, U=UPDATE, Q=QUIT";
FOR B = 1 TO 15                    ' DRAW SLIDE CONTROLS
 IF B = 9 THEN B = 10
 IF B < 9 THEN
  Y = (C(8, B) / 3)
  FOR A = 1 TO 7                        ' DRAW KNOB CONTROLS
CIRCLE (B * 40, A * 40), 12, 7: DRAW "P4,7"
LINE -STEP(-SIN((C(A, B) * .0244)) * 10, COS((C(A, B) * .0244)) * 10), 15
  NEXT A
 ELSE
  Y = (C(B - 7, 9) / 3)
 END IF
 LINE (B * 40 - 15, 422)-(B * 40 + 15, 328), 0, BF
 LINE (B * 40 - 13, 420 - Y)-(B * 40 + 13, 414 - Y), 15, B
 LINE (B * 40 - 12, 419 - Y)-(B * 40 + 12, 415 - Y), 1, BF
NEXT B
PSET ((CHAN * 40 - 25), (KNOB * 40)): DRAW "C0 R5 H3 F3 G3"
key$ = INKEY$: IF key$ = "q" OR key$ = "Q" THEN END
IF key$ = "S" OR key$ = "s" THEN
 LOCATE 29, 1: PRINT "                                                   ";
 LOCATE 29, 1: INPUT ; "FILE NAME TO SAVE AS:", FILEN$
 OPEN FILEN$ FOR OUTPUT AS #1
 FOR A = 0 TO 9
  FOR B = 0 TO 8: WRITE #1, C(B, A): NEXT B
 NEXT A
 CLOSE #1
END IF
IF key$ = "R" OR key$ = "r" THEN
 LOCATE 29, 1: PRINT "                                                   ";
 LOCATE 29, 1: INPUT ; "FILE NAME TO RECALL: ", FILEN$
 OPEN FILEN$ FOR INPUT AS #1
 FOR A = 0 TO 9
  FOR B = 0 TO 8: INPUT #1, C(B, A): NEXT B
 NEXT A
 CLOSE #1
END IF
IF key$ = "U" OR key$ = "u" THEN
 OLDKNOB = KNOB
 OLDCHAN = CHAN
 FOR KNOB = 0 TO 8
  FOR CHAN = 0 TO 9
IF CHAN < 9 THEN OUT DOUT, CHAN + 95 ELSE OUT DOUT, 64
NEWKNOB = KNOB - 2: IF NEWKNOB < 1 THEN NEWKNOB = NEWKNOB + 8
```

```
FOR A = 1 TO 8
 IF A = NEWKNOB THEN OUT COUT, 1: OUT COUT, 0 ELSE OUT COUT, 3: OUT COUT,
  2
NEXT A
FOR A = 7 TO 0 STEP -1
 SETTING = (C(KNOB, CHAN) AND 2 ^ A)
 IF SETTING > 0 THEN OUT COUT, 1: OUT COUT, 0 ELSE OUT COUT, 3: OUT COUT,
  2
NEXT A
OUT DOUT, 255: OUT COUT, 255
  NEXT CHAN
 NEXT KNOB
 KNOB = OLDKNOB
 CHAN = OLDCHAN
END IF
IF key$ = "4" THEN CHAN = CHAN - 1: IF CHAN = 0 THEN CHAN = 9
IF key$ = "6" THEN CHAN = CHAN + 1: IF CHAN = 10 THEN CHAN = 1
IF key$ = "8" THEN KNOB = KNOB - 1: IF KNOB = 0 THEN KNOB = 8
IF key$ = "2" THEN KNOB = KNOB + 1: IF KNOB = 9 THEN KNOB = 1
IF key$ = "-" OR key$ = "_" THEN C(KNOB, CHAN) = C(KNOB, CHAN)- 4
IF key$ = "+" OR key$ = "=" THEN C(KNOB, CHAN) = C(KNOB, CHAN)+ 4
IF C(KNOB, CHAN) < 0 THEN C(KNOB, CHAN) = 252
IF C(KNOB, CHAN) > 252 THEN C(KNOB, CHAN) = 0
PSET ((CHAN * 40 - 25), (KNOB * 40)): DRAW "C15 R5 H3 F3 G3"
LOCATE 29, 65: PRINT "SETTING =   "; : LOCATE 29, 73: PRINT C(KNOB, CHAN);
IF CHAN < 9 THEN OUT DOUT, CHAN + 95 ELSE OUT DOUT, 64
NEWKNOB = KNOB - 2: IF NEWKNOB < 1 THEN NEWKNOB = NEWKNOB + 8
FOR A = 1 TO 8
 IF A = NEWKNOB THEN OUT COUT, 1: OUT COUT, 0 ELSE OUT COUT, 3: OUT COUT,
  2
NEXT A
FOR A = 7 TO 0 STEP -1
 SETTING = (C(KNOB, CHAN) AND 2 ^ A)
 IF SETTING > 0 THEN OUT COUT, 1: OUT COUT, 0 ELSE OUT COUT, 3: OUT COUT,
  2
NEXT A
OUT DOUT, 255: OUT COUT, 255
FOR B = 400 TO 600 STEP 40              ' DRAW VU METERS
 OUT DOUT, (B - 400) / 40               ' SELECT A:D CONVERTER
 LEVEL = (INP(CIN) XOR &H70) \ 16      ' INVERT BIT 7 & SHIFT DATA
 FOR A = 40 TO 280 STEP 16
  CIRCLE (B, A), 5: PAINT (B, A), 0, 15
  IF LEVEL <= (A - 40) / 16 THEN PAINT (B, A), 2, 15
 NEXT A
 OUT DOUT, 255                          ' DE-SELECT A:D CONVERTER
NEXT B
GOTO START
```

Printer Port EEPROM Verifier 2-9

My EPROM copiers have served me well, but they have some drawbacks. How do you determine whether an ultraviolet-erasable EPROM is, in fact, completely erased? How do you determine whether it is, in fact, programmed correctly?

This project can show you that an EPROM is completely erased, and it can also give a checksum to determine proper programming. If everything had worked the way I had hoped, it would have also programmed one. The problem with programming a BIOS is that an "End of File" marker, ASCII 26, is quite often contained in the code to be programmed into a BIOS. Try as I may, Quick BASIC won't let me get past any EOF marker that might be there. Perhaps another programming language such as C would allow you to get past the false "End of File" markers.

Qty	Part
3	74LS393
1	74LS157
1	74LS245
1	74LS14
4	14-pin sockets
1	16-pin socket
1	20-pin socket
1	32-pin socket (ZIF socket optional)
1	5-volt and 12-volt DC-AC adapter
5 feet	20-conductor wire
1	25-pin plug and hood to match
1	circuit board, 2 × 3 inches
1	box, 2 × 3 × 1 inches
Miscellaneous	resistors and capacitors
Optional power supply arrangement:	
1	25-volt center tapped transformer
1	50-volt, 1-amp bridge rectifier
1	7812 voltage regulator
1	7805 voltage regulator
2	470-µF/16-volt capacitors

Table 2-23: Parts list

Data is read at a rate of one-half byte at a time, starting with address zero. A reset on D2 sets all counters to zero, thus selecting address zero. Once again, I included an inverter to correct for the inverted D7 on the printer port.

Three 74LS393 counters provide all addresses needed. In fact, there is room for expansion to larger EPROMs in the future. Two jumpers are available in order to

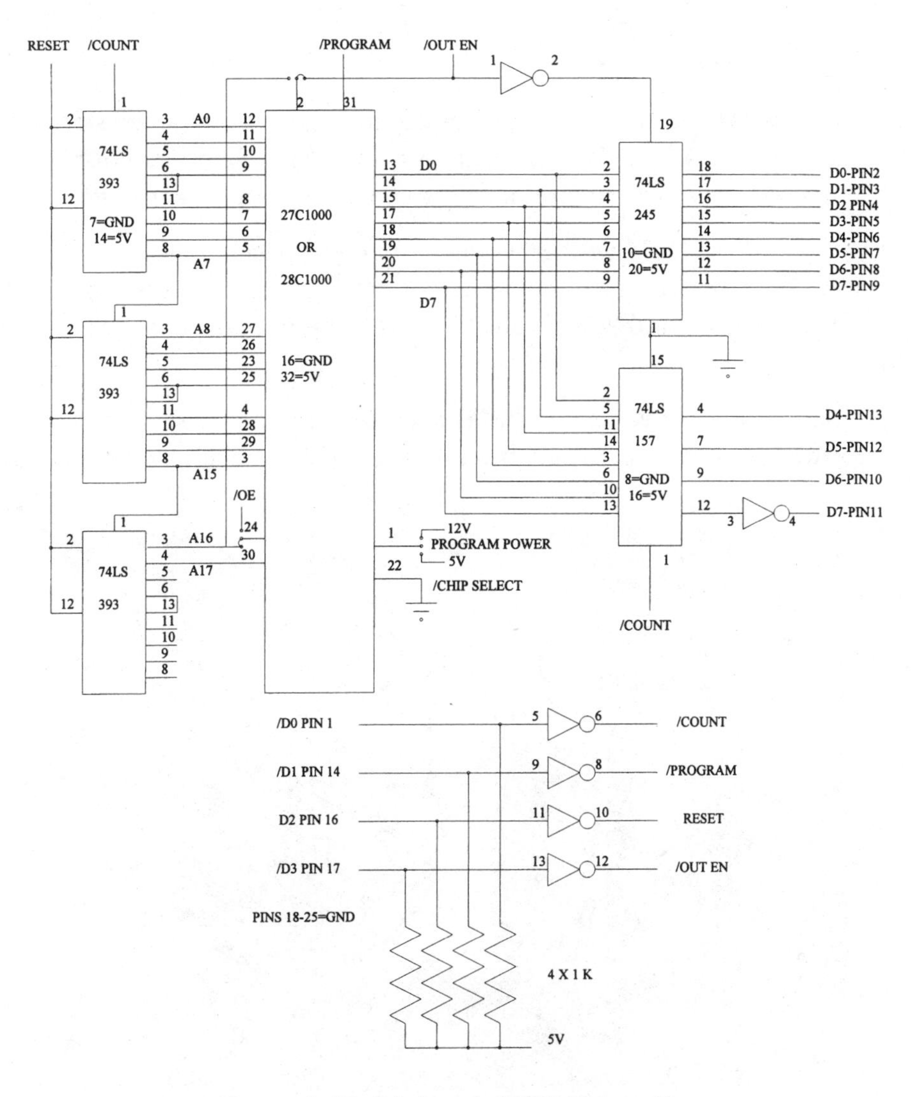

Figure 2-32: Printer port EPROM verifier

swap Output enable and A16 for 28F1000 versus 28F010 pinouts. Another jumper selects between 5 volts and 12 volts. This selects the proper power supplies in order to program the EEPROMs.

As you can see from the schematic, Figure 2-32, the design is rather simple. In this project, only six ICs are used in addition to the EPROM being read. For simplicity, I used an AC adapter that had built-in 5- and 12-volt regulators for this project.

Don't forget the filter capacitors!

The software is in Quick BASIC. It will display the address on the left side of the screen followed by 64 bytes of data. Data bytes less than 32 are not displayed because the screen would interpret them as line feeds, clear screens, etc.

Once you get the EPROM verifier working, you will quickly note that 1-megabit BIOS chips use only the first half of the chip. That makes it possible to interchange 28F1000 and 28F010s without needing to make changes on the motherboard.

The program asks for a file name. This is necessary in order to save the contents of the EPROM for future reference, but the option could be disabled if it is not

Figure 2-33: Picture of the printer port EPROM verifier

wanted. There is also an optional line to reduce the program's speed for easier viewing on the screen.

Upon completion of the program, a checksum value will be displayed. You can record this information and determine if the EPROM matches another, or if it matches the master from which it was copied.

```
' PRINTER PORT EEPROM VERIFIER 6/17/98 BY BOB DAVIS
CLS : LOCATE 10, 20: INPUT "USE PRINTER PORT NUMBER: ", LPT
LOCATE 12, 20: INPUT "FILE NAME TO SAVE CONTENTS TO: ", FILE$
DOUT = &H378: COUT = &H37A: CIN = &H379
IF LPT = 1 THEN DOUT = &H3BC: COUT = &H3BE: CIN = &H3BD
IF LPT = 3 THEN DOUT = &H278: COUT = &H27A: CIN = &H279
OPEN FILE$ FOR OUTPUT AS #1
CLS
OUT COUT, 3                                 'RESET - START
START:
DISP = 0
CSUM = 0
FOR ADX = 1 TO 131072
OUT COUT, 6
D = (INP(CIN) \ 16)                     'INPUT AND RIGHT SHIFT
OUT COUT, 7
D = D + (INP(CIN) AND &HF0)            'INPUT HIGH 1/2 BYTE
DISP = DISP + 1
IF DISP = 64 THEN DISP = 0: PRINT "": PRINT ADX;
IF D > 32 THEN PRINT CHR$(D);  ELSE PRINT " ";         'DISPLAY ON SCREEN
PRINT #1, CHR$(D);                      'SAVE TO THE FILE
CSUM = CSUM + D
  ' FOR A = 1 TO 100: NEXT A                'OPTIONAL SLOW FOR VIEWING
NEXT ADX
PRINT "": PRINT "CHECKSUM ="; : PRINT CSUM
CLOSE #1
```

Chapter 3

Monitor Projects

In this chapter you will discover:

These monitor modifications started way back in the early 1980s with surplus black-and-white monitors. They were being converted to Mono, Hercules, and CGA compatibility and vice versa. Those monitors sometimes didn't even have a cabinet. Those were the days! These days, anything less than Color VGA is junk, and LCD displays are dropping in price to where they are approaching the cost of a good monitor.

Around 1987 I purchased an "EGA" monitor without a cabinet or schematics for only $175. At that time that, this was a great price for a color monitor, even if it didn't have a cabinet. The catches on this deal included the problem that it only worked at 24 kHz. Adapter cards were available, at the time, but were very expensive. Some people bought this surplus monitor, as well as others like it, and were forced to pay top dollar for the adapter card. So instead, some converted it to true EGA. First mine was converted to 22 kHz and then, by dropping the B+ voltage and making some other changes, it was converted to both 22 kHz and 16 kHz.

Some time later, two attempts were made to convert the 24-kHz monitor to VGA compatibility. Both had limited success and typically resulted in having an unrepairable or dead monitor. When the monitor did work in VGA mode, the picture was dim, and a character or two were lost off the side of the picture. It was my first conversion to VGA, and I admit it didn't work very well.

The valuable lessons I learned were applied to many more successful conversions, first converting 30-kHz IBM monitors to VGA, and then converting 26-kHz IBM monitors. They sometimes had a dimmer and narrower picture after conversion, but they proved to work quite well. Their picture quality improved dramatically with the addition of analog video amplifiers. I must have converted several dozen of these monitors over the years.

During that time, I also purchased some surplus NEC monitors. They took little more than an adapter cable and a switch position change to be easily converted to VGA. Automatic vertical size was first developed for those NEC monitors, and that helped to improve their performance.

This chapter also includes the conversion of the Hitachi H-4119, a 64-kHz monitor. Initially, I purchased three monitors and converted them to VGA and SVGA compatibility. One of those monitors is still in use today, and I was able to sell the others. Since then, I've converted more than a dozen with practically no problems. I did fry one or two and had to replace the horizontal output transistors. When buying them, watch out for CRTs that are so worn down that the picture is quite dim. A possible fix for a dim picture is to add a 6-volt power supply to give a boost to the CRT filament. These converted monitors sometimes are a little weak on the filament voltage. Now I'm working on an adapter to convert VGA to 63 kHz and this will avoid the need to modify the Hitachi monitors.

Automatic Monitor Shutoff 3-1

For years (in vain), I've hung signs saying "Turn off the monitor when it is not in use." What software companies call "screen savers" do little more than entertain people that are passing by. Something needed to be done. So I developed this simple control box. It watches the keyboard's data line and shuts off the monitor during long periods of inactivity. Simply press a key on the keyboard and the monitor comes back on. It will increase the life of a monitor left unattended and will also save on the electric bill.

The circuit uses a 556 dual timer. These timers are set up as dual one-shot multivibrators called single- or one-shots. The first one is used as a pulse stretcher, this keeps the capacitor of the second one discharged. Whenever a key is pressed, it shorts the capacitor to ground. Both single-shots are wired to trigger every time a key is pressed. In this manner, the circuit won't time out unless there is a long period of inactivity (about 10 minutes).

The 12-volt power source can be provided by way of an AC adapter or a 110-volt to a 9 to 12 AC volt transformer. The transformer will then need a 50-volt, 1-amp bridge rectifier and 500-μF, 16-volt filter capacitor. You might have to change the 100-ohm resistor to attain a value of 9 volts for the 556. The output relay is a Radio Shack 12 VDC relay that can handle up to 10 amps. The 100-ohm resistor in series with the power source may also need to be increased in order to reduce the sensitivity to noise from the power source and from the keyboard cable.

The control box I used was only 6 inches by 3 inches by 2 inches deep, and it had adequate room for an internal transformer. The 110 volts is provided by a power cord with a molded three-prong plug. It then powers the monitor via a 110-volt three-prong socket. The keyboard socket is one with earlike connectors that screw onto the side of the control box. You can use the plug from a discarded keyboard or keyboard extension cable—just cut one end off.

1	556 timer
1	14-pin socket
2	2N2222 transistors
1	12-volt, 10-amp relay
1	120-volt to 9-volt transformer
1	bridge rectifier
1	keyboard cable
1	keyboard jack
1	120-volt power cord
1	120-volt outlet
1	circuit board, 2 × 2 inches
1	box, 3 × 5 × 2 inches
Miscellaneous	resistors and capacitors

Table 3-1: Parts list

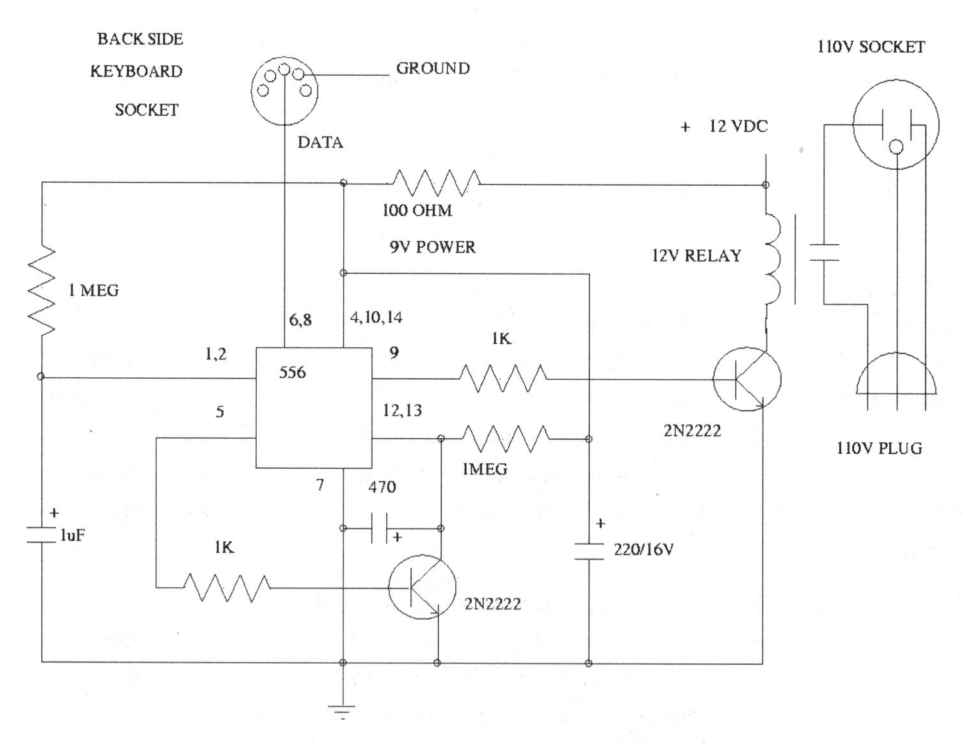

Figure 3-1: Automatic monitor switch

VGA-to-CGA Adapter 3-2

Adapters used to convert VGA to be viewed on a television have been around for a while. For $150 to $250 you can get a pretty good picture on a television set. This is fine if you're a salesman and need a large-screen display. But if you're a hobbyist like me, it's not worth the cost. Then again, you may not need all of the fancy converter features, such as anti-flicker and 64,000 or more colors. If this is the case, an adapter can be built for about $40.

Professional adapters are expensive because each primary color, red, green, and blue, must be converted from analog back to digital. It can then be stored into memory and recalled at a different rate. Then, it is converted back to analog again. This requires three 30-MHz analog-to-digital converters, 256K or more of high-speed memory, and three 15-MHz digital-to-analog converters. It's no wonder they cost so much!

A different video rate is necessary because television operates at 15,750 scans per second. It then draws out an entire picture in 1/30 of a second. VGA operates twice as fast. VGA scans at 31,500 cycles per second and draws a picture in 1/60 of a second. VGA also varies the time to complete a picture between 1/60 and 1/72 of a second, depending on the VGA mode selected and the graphics card. This may result in a rolling picture in some modes on some TVs when using this converter.

Television also "overscans" its picture—that is, the picture continues off the top, bottom, and sides of the picture. However VGA "underscans" its picture—that is, the picture ends just short of the edges on the screen. This results in some information being lost on the edges of a television when using this converter.

If you try to display VGA on a TV set without using a full converter—and it can be done—you get TWO pictures. VGA, because it is working at 60 pictures per second, completes two pictures in 1/30 of a second. Each picture is half of its normal width. Text is so squeezed together that it is usually illegible. It is like trying to fit 160 columns of text on every line of the screen. To solve the problem, the VGA picture has to be fed into memory and then recalled at half the rate at which it was entered.

The problem is that information coming out of the VGA adapter's memory was already converted to analog. It must be converted back to digital so it can again be stored into memory. To avoid expensive converters, a two-bit flash converter will be used for each color. It works by raising the video approximately 0.2 to 0.6 volts so that it will trigger a LS TTL input. LS TTL requires 0.8 to 1.2 volts to operate properly, and the video is normally 0.6 to 0.8 volts in amplitude. The drawback of this method of conversion is that it doesn't handle shades of colors very well. These quite often come out as "sparkles" because they exist on the borderline. A 74LS367 IC was used for this conversion. To reduce the amount of sparkles, use two filter capacitors on the power and ground pins.

The next problem is the amount of memory necessary. In order to convert an entire picture, it must first be stored into memory. This would require 256K or more of memory. But if the picture was processed one line at a time, only 1K of memory is needed. The drawback of this method is that every other line will be lost. This can result in some flicker and loss of clarity. On the up side, we considerably reduced the price and quantity of parts necessary.

1	UPD42102C-5 (available from Marshall Industries 9320 Telstar Avenue, El Monte, CA 91731)
1	74LS367
1	74LS86
1	74LS02
1	MC1377P1
1	LM7805
1	16-MHz clock
1	30-MHz clock
1	3.58-MHz crystal
4	14-pin sockets
1	20-pin socket
1	24-pin socket
1	22 µH coil (coils are available from Newark Electronics)
1	39 µH coil
1	100 µH coil
1	9 VDC-AC adapter
1	LM7805 voltage regulator
1	box, 4 × 3 × 1.5 inches
1	circuit board, 4 × 3 inches
Miscellaneous	resistors, capacitors, etc.

Table 3-2: Parts list

NEC makes a chip that can be used as a video line buffer. The chip is dual ported—therefore, you can write into it and read out of it at two different rates. It is the UPD42102C-5, available from Marshall Industries.

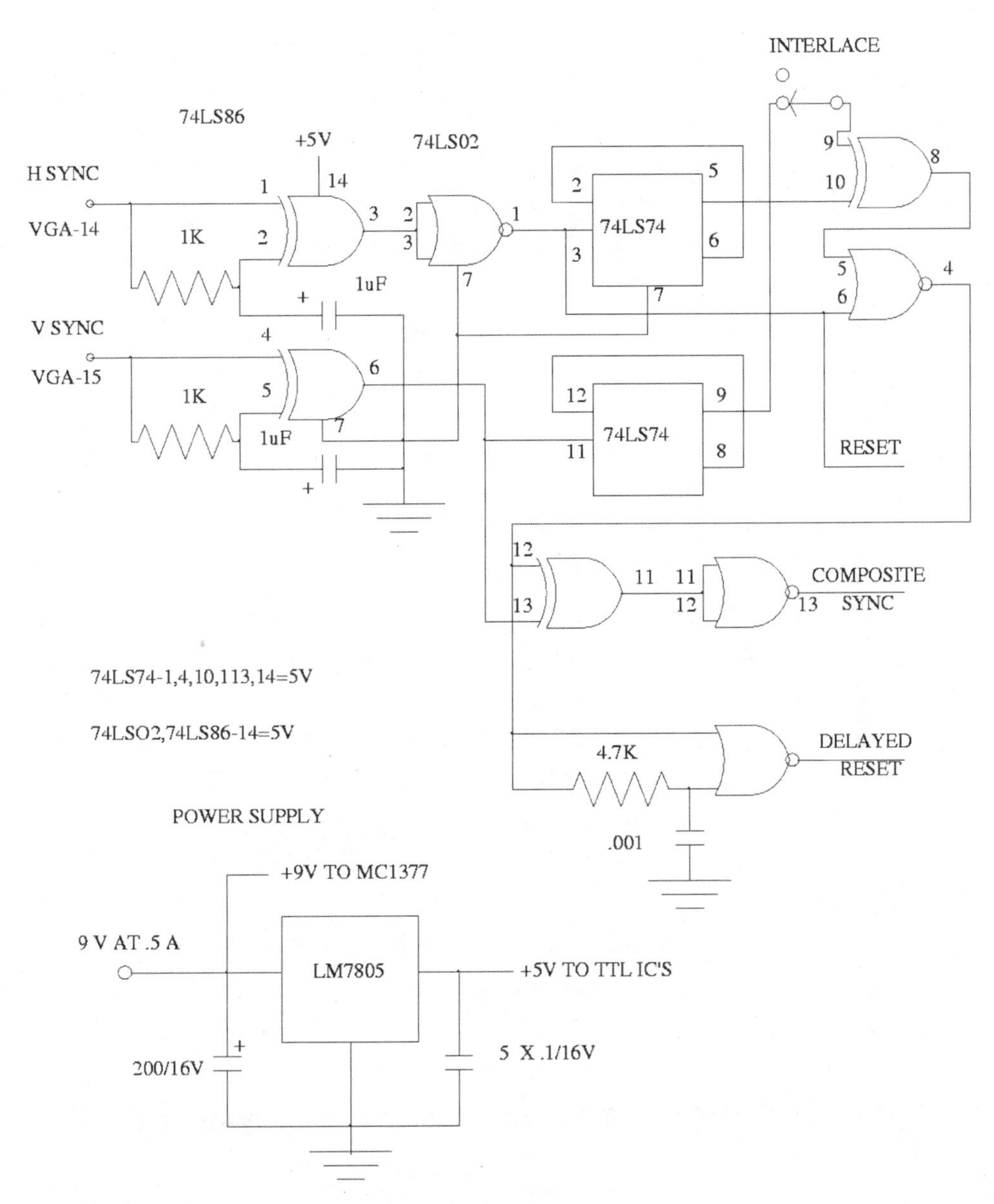

Figure 3-2: VGA-to-CGA and TV adapter control section

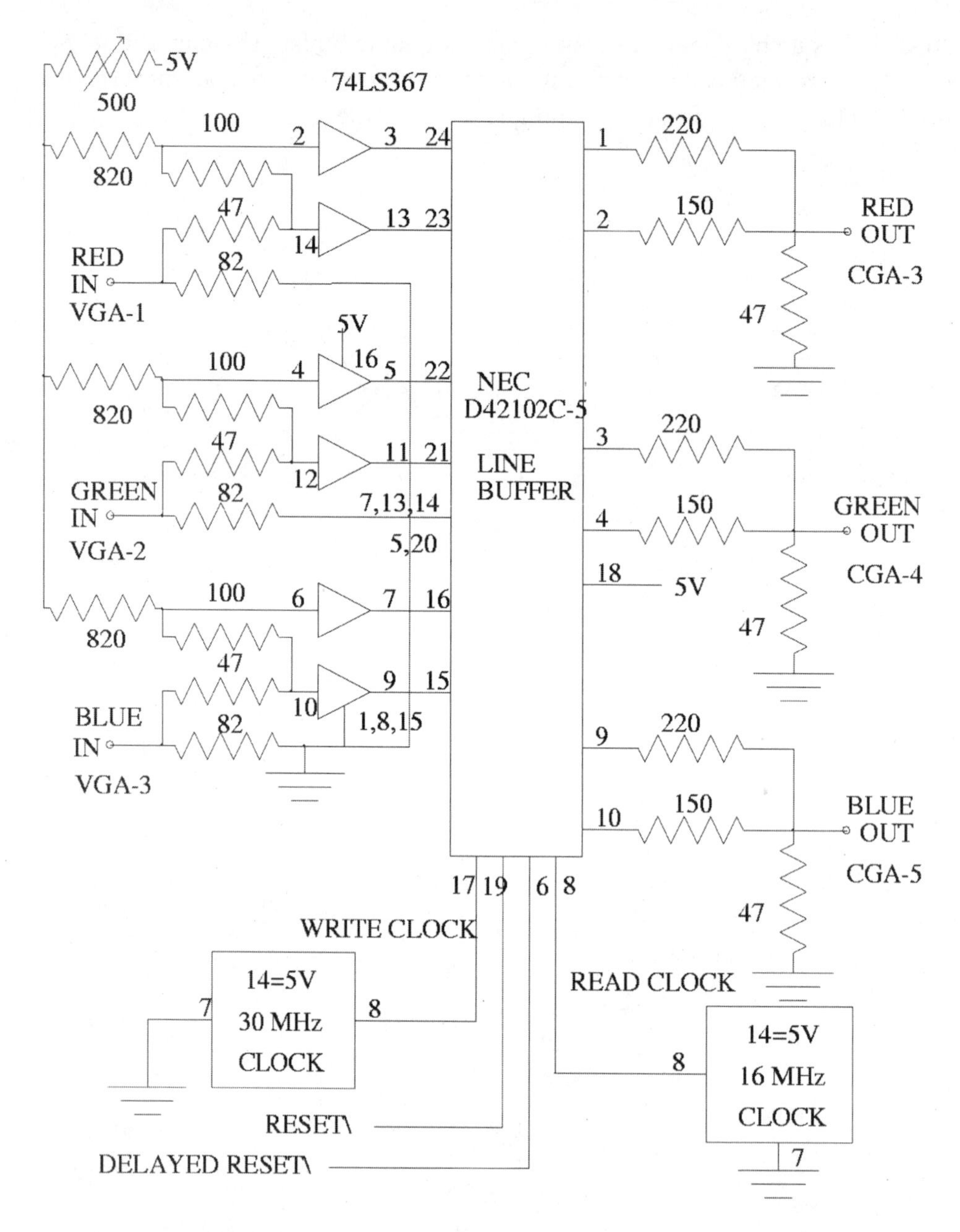

Figure 3-3: VGA-to-CGA and TV adapter memory section

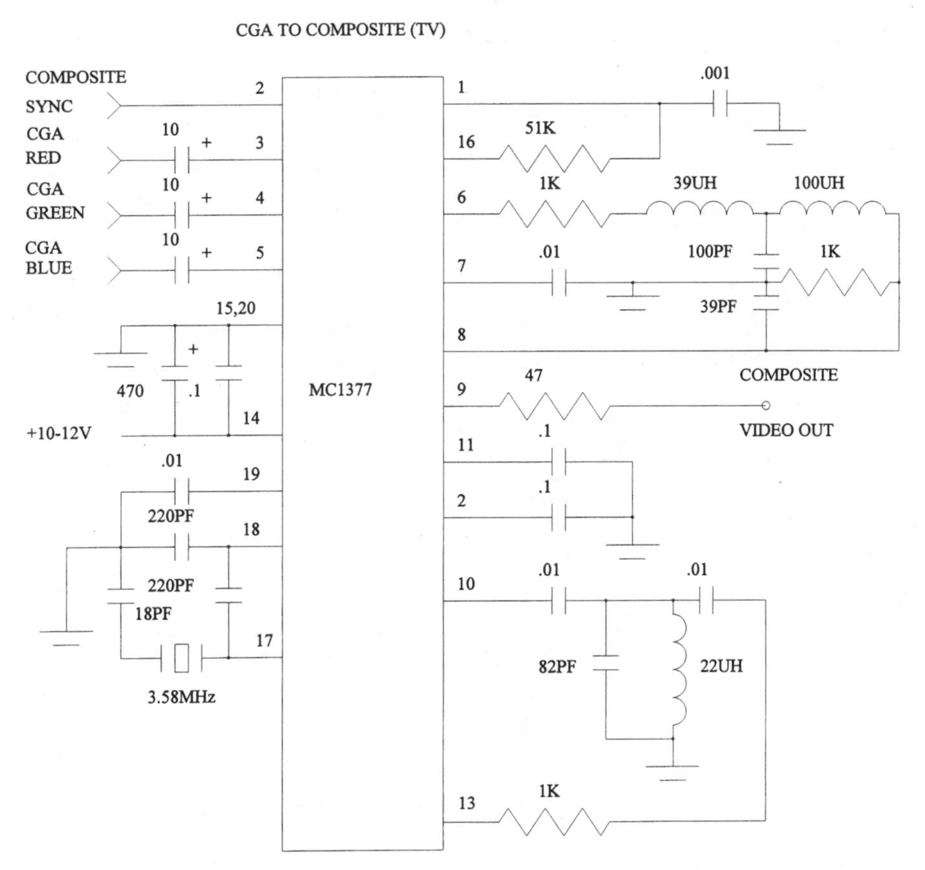

Figure 3-4: VGA-to-CGA and TV adapter composite section

The next step is to convert the digital back to analog. Here three resistors will do the job. This will not be very precise, but it is much easier than using digital-to-analog converters.

The resultant three colors must be encoded together into the NTSC format for television. Here the encoder is easily available on one chip, the MC1377P. The output is composite NTSC video. The TV must have a video input to use this signal. Otherwise, a VCR will work, as VCRs almost always have a video input.

Also needed for this converter to work is a 74LS86 sync phase correction circuit. VGA changes the phase and this signals the monitor that it is in different modes. Once this is corrected, it is divided by 2 in a 74LS74 to get half the rate. It then must be combined into composite sync for NTSC compatibility.

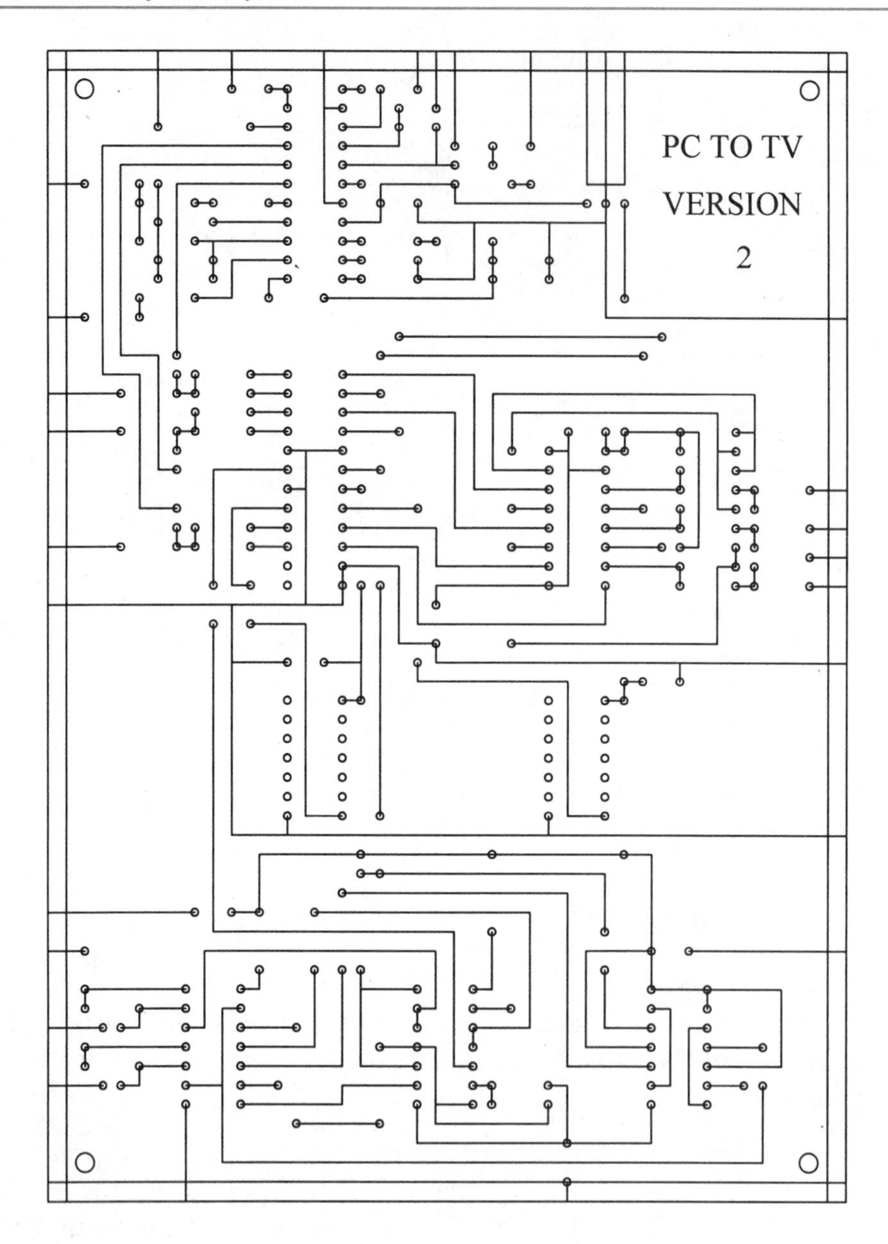

Figure 3-5: Circuit board

The master clock source is a 30-MHz clock oscillator. It is also divided by 2 in the 74LS74 to get the read clock. Since the information is written at 30 MHz and read at 15 MHz, it comes out at half the rate.

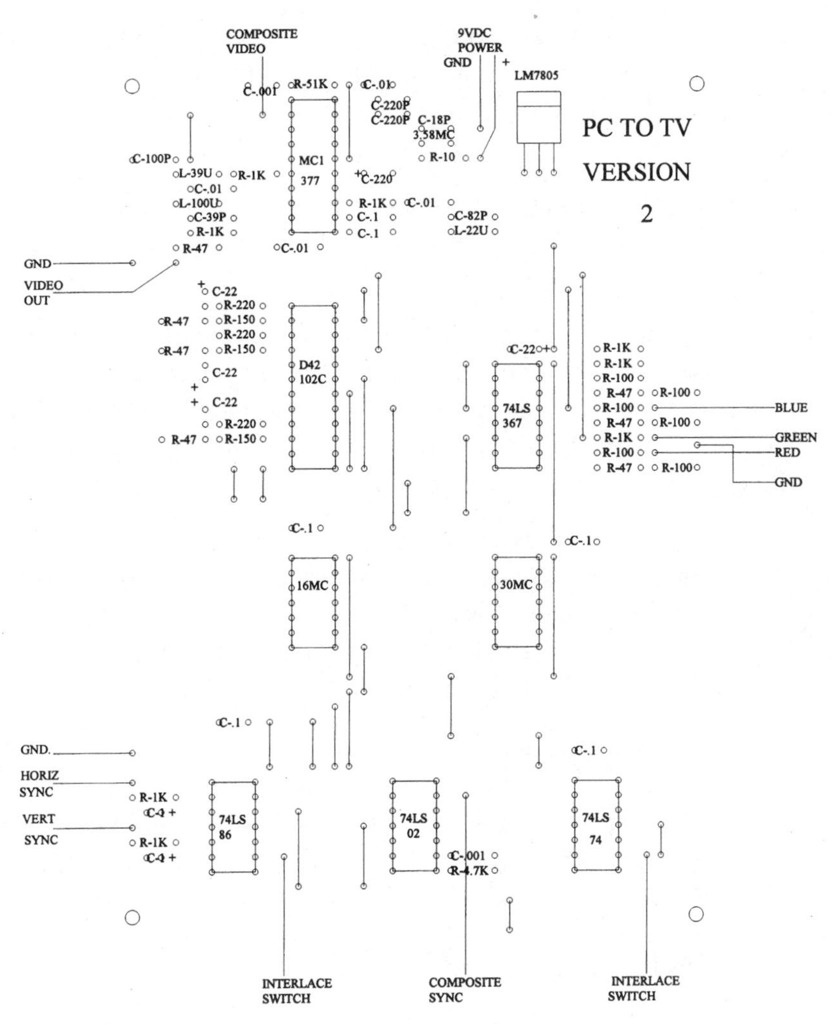

Figure 3-6: Parts layout

Power comes from an AC adapter providing 9 to 12 volts DC. It powers the MC1377 directly, but it goes through a 7805 regulator to power the TTL ICs. No heat sink is required because the circuit draws very little power.

The schematic is divided into three drawings. First is the analog-to-digital converters, memory, and digital-to-analog converters. Second is the sync and control circuits. Third is the NTSC encoder schematic. The schematic doesn't show outputs for connecting a CGA adapter. A CGA monitor will give a better picture than a TV set. Pins 11 and 12 of the 74LS86 would supply sync signals for such a CGA monitor.

The converter can fit into a box about 4 × 3 × 1.5 inches. Make sure to leave lots of room for all the coils and capacitors that are needed around the MC1377 for it to work. The MC1377 design is modified slightly from the normal. This avoids using some of the proprietary coils normally associated with it.

The circuit is assembled on the circuit board that came with the box. Wire wrap is used on all but the MC1377, which is in a socket and connected directly to all of the necessary support components.

For testing purposes, a CGA monitor can be used. An o'scope or logic probe will work if you are good at wiring and troubleshooting. Check the sync signals before connecting a monitor. Check the analog-to-digital converters for a toggling output, as well as the master clock and clock divided by 2. Then, check the outputs of the line buffer and connect the monitor. If it works to this point, try the composite output to a VCR and TV set.

On the up side, this adapter works and the picture is quite readable. On the down side, the picture is a little jagged. The circuit, when used with a TV set, is a little difficult to read. The poor readability is a limitation in the video bandwidth of a TV set. It is limited to about 4.5 MHz for black and white, and the VGA output is 28 MHz, which is converted to 14 MHz. Also, television requires the color be encoded at 3.58 MHz for a color picture.

In 640 by 400 mode the picture rolls vertically on some newer TVs because of the 70-Hz vertical sync. This isn't a problem in the 640 by 480 mode that Windows 3.1 uses, because that changes the vertical to 60 Hz. However, CGA-compatible monitors don't have those drawbacks, and some even have direct analog inputs, as well as 50-MHz video response.

VGA-to-CGA Adapter: Simplified 3-3

The process of making an adapter that converts from one monitor standard to another can be a lot easier than you think. It is simple as long as the change is to either double or half the original horizontal frequency—that, and the vertical frequency must be kept the same. All that is needed for this converter is a dual ported video line buffer IC, a clock, and a couple of support ICs.

A few years back, I developed an 8 IC VGA-to-CGA converter. This adapter could even produce composite analog video for connection to a TV set. However, in my never-ending quest to simplify it, it has now been successfully reduced to only 4 ICs and a couple of diodes. Since it converts VGA to CGA, a lot of the normal conversion circuitry is not necessary.

A full VGA-to-TV converter requires several parts. First, there are three high-speed analog-to-digital converters. Also required is a high-speed dual port video memory for each color, with enough memory addresses to hold an entire frame of the picture. Next there have to be three 15-MHz digital-to-analog converters to convert each color back to analog at CGA or TV rates. The control section has to handle interlacing of the video and multiple horizontal and vertical frequencies. On top of that, there is a converter to change the CGA signals to NTSC for connection to a TV set.

Since this converter is changing VGA to CGA, it only needs an on/off signal for each color. CGA does not recognize analog levels. It only uses TTL (transistor transistor logic) levels. This means that no analog-to-digital or digital-to-analog converters will be needed.

VGA levels are usually insufficient in amplitude to drive a TTL circuit. Normally, an amplifier or comparator is used to increase this level. Adding 1K pull-up resistors will bias the normally 0- to 1-volt signal to the 1- to 2-volt level needed to drive logic levels. It will, however, allow some noise to be displayed if the analog level falls in the "gray" area of neither a 1 nor a 0. Adding a Schmidt trigger such as a 74LS14 will reduce the noise.

By not allowing the conversion of vertical frequencies between 60 and 70 Hz, the memory can be reduced to about 1K of dual ported memory. That is enough memory to hold one horizontal line of 640 pixels with room to spare. This means that whatever vertical frequency you start with, it will remain the same after conversion. Most CGA monitors have no problem with this, especially if they have a vertical hold control. Some TVs will not work as well because they cannot lock onto VGA's 70-Hz vertical.

In order to convert VGA analog levels to TTL levels, some resistors are used. These resistors bias the signal at a high enough level to trigger the inputs of the memory chip. The memory IC is a NEC D42102C-5, available from Marshall Industries. It contains two counters with independent count and reset signals for each. The counters then select the desired memory addresses via dual ports, and each counter can access a different address at the same time.

In converting down to half the frequency, the converter writes a horizontal line into memory at twice the speed of the read function. Since half the data will never be read, it is discarded. However, it still shows up on the next frame because of interlacing. This can cause some horizontal lines to flicker, since they only show up on every other picture or frame.

The control section uses one-half of a 74HCT86 to correct for phase differences in the many VGA modes of operation. Then one-half of a 74ALS74 divides the horizontal sync by 2, because we only need every other one. Two diodes form a gate in order to get one sync pulse for every two coming in. The other half of the 74ALS74 divides the clock by 2 so the converter can read out at one-half the rate at which it writes. The second half of the 74HCT86 forms, and inverts, a composite horizontal sync. This is not needed for CGA, but is needed for composite monitors. This is also necessary to feed an MC1377 in order to convert the CGA output to a composite color television signal (NTSC).

In order to convert the CGA output to composite video and feed a VCR and a normal TV set, just add a MC1377 converter. A design for one of these can be

Quantity	Part Number	Source
	UPD42102C-5	Marshall Industries, 9320 Telstar Ave., El Monte, CA 91731
1	74HCT86 (can substitute a 74LS86 with 1K resistors and 1 µF caps)	Mouser, 1-800-346-6873.
1	74ALS74 (can substitute a 74LS74)	Radio Shack or Mouser, 1-800-346-6873.
1	LM7805	Radio Shack or Mouser, 1-800-346-6873.
1	32-MHz clock	Mouser, 1-800-346-6873.
3	14-pin IC sockets	
1	24-pin IC socket	
2	1N4148 diodes	
1	9 VDC–AC adapter	
1	box, 3 × 2 × 1 inches	
1	circuit board, 3 × 2 inches	
Miscellaneous	resistors and capacitors	Radio Shack

Table 3-3: Parts sources

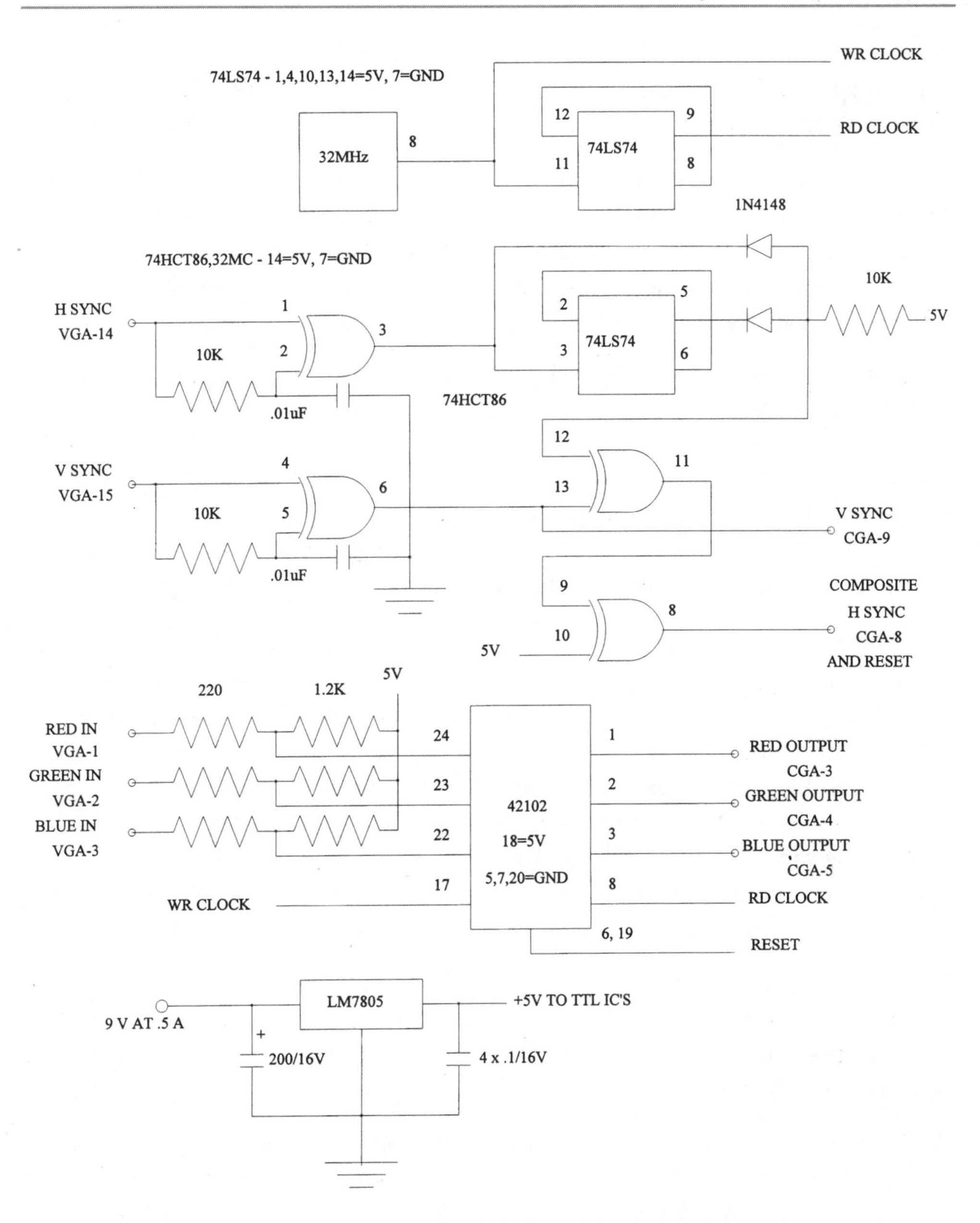

Figure 3-7: VGA-to-CGA adapter schematic

found in the April 1995 edition of *Nuts and Volts*, on page 113. The MC1377 requires special coils for operation, which are hard to find, as you'll learn in that project. A simpler version can be found in my older VGA-to-TV converter design.

Vertical lines in the converted picture will show some jagged spots. This is because the clock generator is not phase locked to the sync signals. One possible solution would be to use a phase-locked loop. The problem here is that most phase-locked loops will not operate at the needed 32-MHz frequency. One other alternative is to tap into the clock frequency of the VGA card at its "feature" connector.

This converter, with a few simple changes, can work backwards in order to convert CGA up to VGA. First, swap the read and write clocks so that it will read twice as fast as it writes into memory. That means that the first line after the horizontal sync will be a repeat of the previous line. The next line will then be the current line. Leave out the half of the 74ALS74 used for dividing the horizontal sync by 2, as well as the two diode gate. The 74HCT86 pin 3 connects directly to pin 12. This way, the converter can be used to convert CGA to VGA.

A backwards converter can convert VGA to be used with a 65-kHz monitor such as a Hitachi HM-4119. It doesn't work sufficiently when converting VGA to 65 kHz. First of all, the monitor will be missing every other horizontal sync pulse. With most monitors, that is not a serious problem. Second, the sample rate will be only one-half the sampled frequency. But in order to properly sample a signal, the sample rate must exceed the sampled signal! When converting from VGA to 65 kHz, the sample rate is only 16 MHz when it should exceed 25 MHz. This is due to the maximum frequency of the memory chip, which is limited by the 32-MHz read rate. So the quality in this mode of operation is somewhat poor.

To get around this, it is possible to use two memory chips. Each chip operates on opposite phases. The problem with this is that the noise levels are too high for any simple design or component layout. The operating speed in this mode is equivalent to 64 MHz!

The simplicity of this converter has been kept a secret for years, but now the "cat is out of the bag." This unit can be made in a box smaller than 2 × 3 inches in size. Point-to-point soldering works fine, but be sure to run a ground wire around the edge of the board first. Mount the 15-pin VGA connector on one end of the box and the 9-pin CGA on the other. Any 9- to 12-volt AC to DC adapter will power the adapter with the regulator shown in the schematic, Figure 3-7. Put a jack on the box for the power input, too. Have fun!

Sync Phase Adapters 3-4

Mono graphics, color graphics, and Hercules graphics always use a fixed phase for their sync signal. VGA, on the other hand, varies the phase to tell the monitor what vertical size to use, as well as what mode the graphics card is in. This 74LS86 design, from its original form up to and including today's version, can correct this problem.

The basic idea behind using the "exclusive OR" gate is that the phase of its output can be controlled by using one of the other inputs. That is to say, it operates as either an inverter or a buffer (i.e., a noninverter), with a control signal.

The sync phase can be positive or negative, but the sync signal is present for only a fraction of the time. The sync signal voltage is either positive or negative. It is in the opposite polarity of the sync phase. When the sync pulses are filtered out with a capacitor, what remains is a voltage that represents the opposite of the signal phase. If this voltage is positive, then the 74LS86 will invert the sync phase. If it is negative, the signal will not be inverted. This results in a guaranteed positive sync phase.

The remaining two gates can be used to mix the two signals for composite sync. They can also be used as inverters by connecting the unused input to 5 volts. The 1K resistor biases the TTL gate; however, a 680-ohm to 1.2K resistor will do. The capacitor value determines how fast it will change with sync phase changes. A value of 1 to 10 μF at 16 volts works fine. The voltage on the capacitor can also be sent to a 74LS139, which can then automatically adjust the verti-

Quantity	Part Number	Source
1	74LS86 or 75HC86	
1	14-pin IC socket	
2	1N914 diodes	
1	box, 2 × 2 × 1 inches	Mouser
1	circuit board, 1 × 1 inch	
3	BNC jacks	Mouser
1	BNC to VGA 15-pin cable	HSC Electronics 1-800-442-5833
or 1	15-pin HD D-sub, male	Mouser
Miscellaneous	resistors and capacitors	

Table 3-4: Parts list

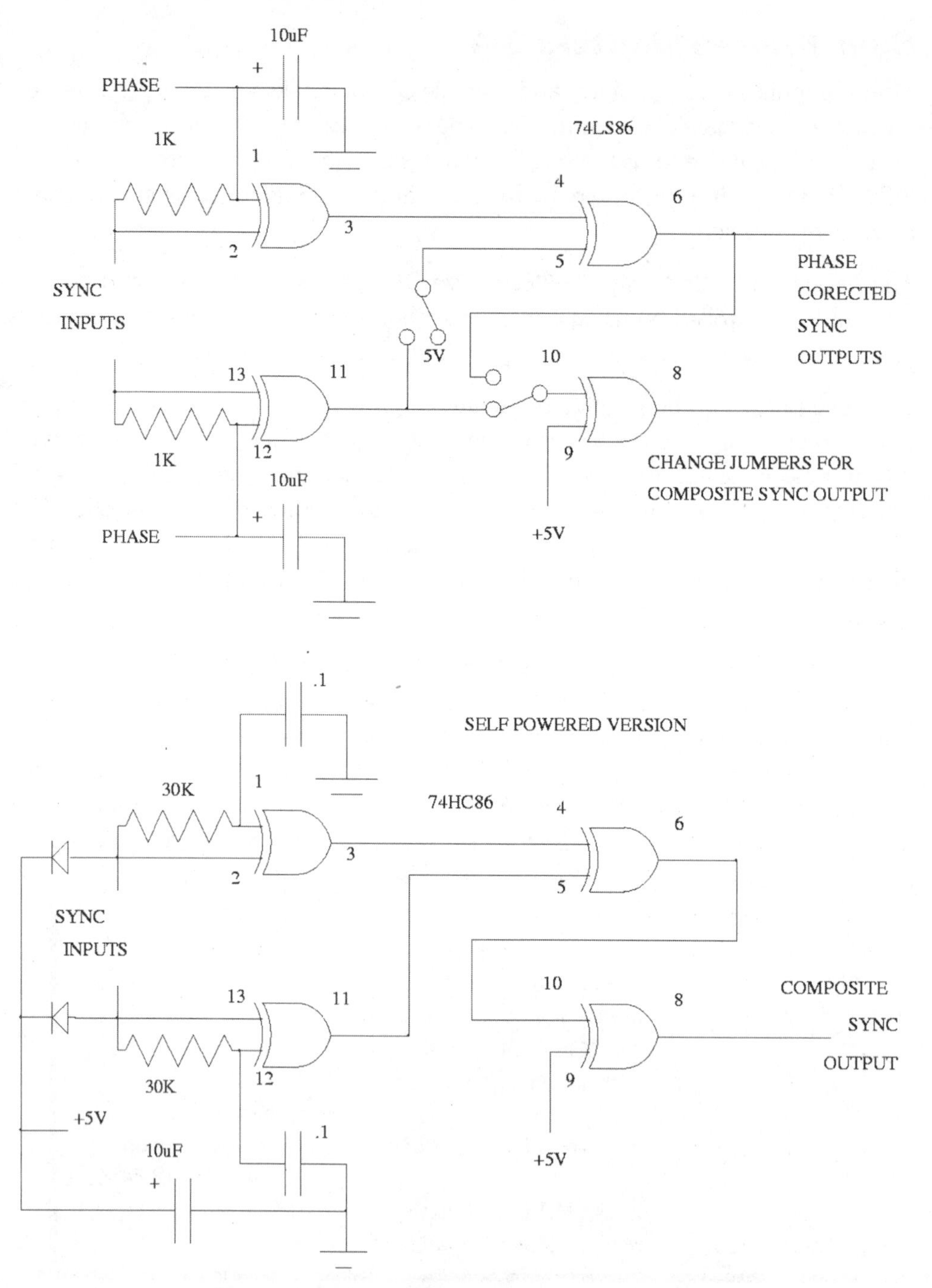

Figure 3-8: Sync phase correction circuits

tor can also be sent to a 74LS139, which can then automatically adjust the vertical size for the four different possibilities.

Recently, a problem with the 74LS86 circuit was solved. It needs 5 volts to operate, so it had to tap into a power supply at some point. The solution is to use a 74HC86. Being a CMOS device, it doesn't need as much power. The source of power (about 4 volts) is obtained via two diodes from the sync inputs. Because the CMOS inputs don't need as much bias, the resistors are increased to about 27 to 33K. The capacitors are reduced to about 0.1 µF. This works because one of the two sync signals is almost always positive, while providing a negative sync pulse. Two diodes configured as an "OR" gate connect them to a power source for the 74HC86.

For a monitor with BNC inputs, use a 15-pin high-density connector to 5 BNC cable available from BG Micro or Lyben. The sync phase adapter can easily fit on a 1-inch square circuit board. It is then installed in a 2 × 2 × 1-inch aluminum box. The box has three BNC jacks—two are on one side for the inputs and one is on the opposite side as the output. Wires from the jacks are used to hold the circuit board in place.

Automatic Vertical Size Circuits 3-5

In order to add an automatic vertical size adjustment to a monitor such as a NEC 1401P3A, just use this simple one-transistor circuit. All that is needed is a circuit to determine if both sync phases are negative. That means that the average voltage will be positive. Use a couple of resistors to charge a capacitor, and then add a transistor to drive a relay. This is shown as the first drawing in Figure 3-9.

On the NEC 1401 monitor, the connector on the control panel pin M4 is 24 volts. This makes a good power source. The relay would connect a 500-ohm

Quantity	Part Number	Source
1	1 × 1″ circuit board	Mouser
1	2N222	MCM Electronics
or 1	74LS3	Mouser
Miscellaneous	resistors and capacitors	

Table 3-5: Parts sources

trimmer potentiometer between pins M3 and M2 and automatically adjust the vertical size.

Referring back to the 74LS86 phase correction circuit, this circuit has capacitors whose voltage is proportional and opposite to the sync phase. This was marked as "phase" in the schematic. A "480 mode" output can be obtained with only one additional IC such as the 74LS38 quad two input gate with open collector output. This IC may already reside in your monitor. In a later section of this chapter you will learn how to remove it, in order to add an analog video amplifier.

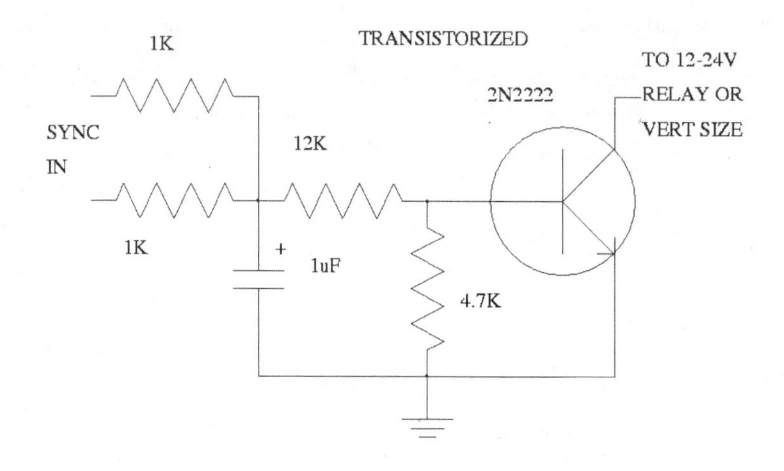

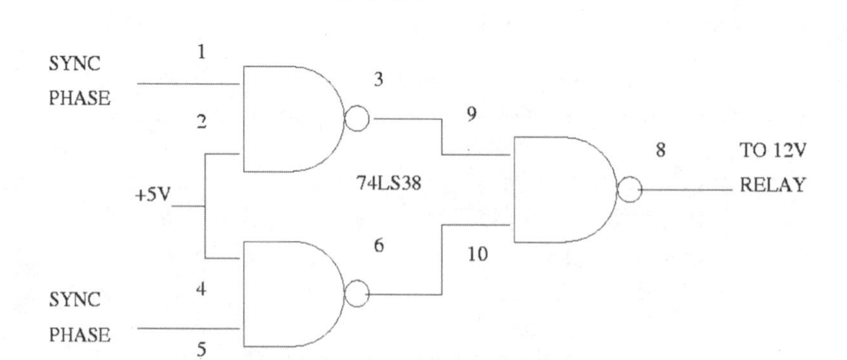

SYNC PHASE INPUTS ARE FROM 74LS86 CIRCUIT

Figure 3-9: Automatic vertical size circuits

The 74LS38 can be set up as two inverters driving an "AND" gate. The gate then directly drives a small 9- or 12-volt relay. The relay contacts when energized, add a trimmer in order to adjust the vertical size control.

In an IBM 3179 monitor, the vertical size relay could be used to add a 5K variable resistor across the vertical size control. It also adds a 1K resistor across the 330-ohm resistor (R421) in series with the vertical position control. Power to the relay can be provided by way of the red wire on the power connector to the video section or any other 12-volt source.

For more control, a 74LS138 can be used. Its inputs are taken from the two "phase" capacitors. It provides an output for each of the four possible sync phase combinations. The drawback is that it can't drive a relay or any circuit that exceeds 5 volts. In order to drive a relay, you can use a ULN2003 or the standard 2N2222s typically used in the other circuits.

If using transistors to drive the relays, don't forget the 4.7K to 10K resistors in series with the base. Also connect the emitter to ground. The ULN2003 IC or transistors can sometimes drive the vertical size control directly, or they can drive relays in order to provide multiple automatic adjustments.

Some monitors need the vertical position and the size adjusted whenever the VGA mode changes. The 74LS138 design allows automatic adjustment in all three VGA modes. It also provides a signal that tells you that the graphics card is in 48-kHz SVGA mode when used with a Trident and some other graphics cards. That signal can drive a relay in order to make automatic horizontal adjustments, or even switch to a different horizontal frequency.

Video Amplifiers 3-6

The simplest method of working with VGA analog video is to convert it to TTL. "LS" TTL ICs have an input that is almost analog compatible and works well for

Quantity	Part Number	Source
3	LM733 or NE592	MCM Electronics
or		
1	74LS04	Mouser
3	2N2369	

Table 3-6: Parts sources

this purpose. Sometimes a bias circuit is needed to help the analog signal trigger the TTL input. This raises the 0.6-volt VGA signal to a higher level to overcome the 1.2- to 1.6-volt TTL input. If the existing circuit in the monitor uses an open collector driver IC such as the 7438, be sure to provide the 74LS04 with the necessary input-level adjusting resistors.

In practical terms, the only analog amplifier circuits I would recommend today would be either the LM733 or NE592 ICs. They work up to about 100 MHz, are cheap, and are readily available. Previously I had used a two- or three-transistor

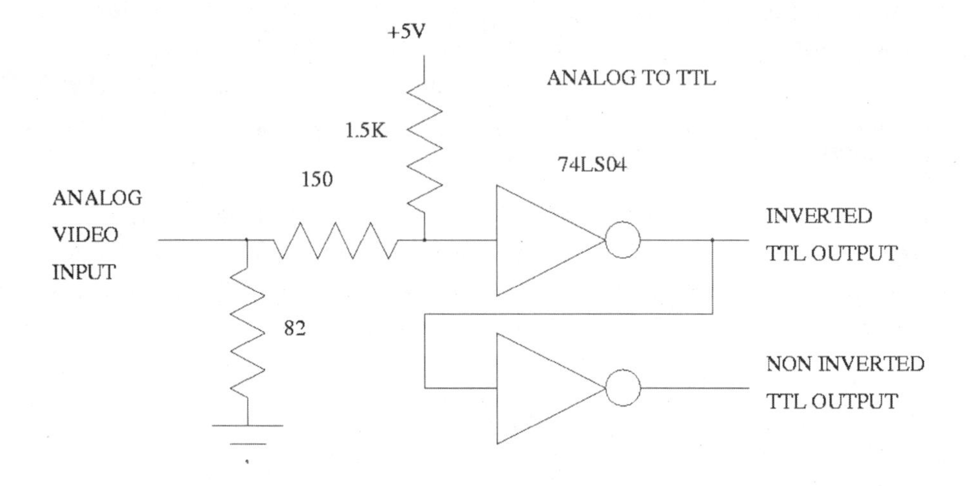

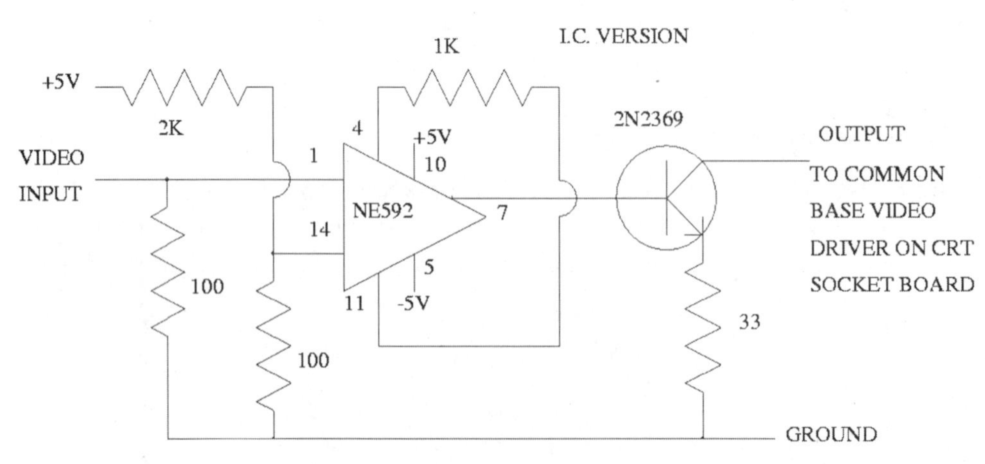

Figure 3-10: Video amplifier circuits

circuit, but that involved too much time and energy. The two ICs operate about the same, except the NE592 needs a 1K resistor to set its gain to 10. That resistor is not needed with the LM733.

The drawback of using ICs is that they require +6 volts and –6 volts to operate. It is often difficult to find a –6-volts source within many monitors. Transistorized amplifiers, on the other hand, usually don't require a negative power supply.

The schematic shown, Figure 3-10, includes a transistor that becomes necessary when replacing open collector TTL driver ICs like the 7438. Along with the transistor, it can drive some CRT socket boards directly. The 2N2369's output connects to the emitter of a common base driver transistor. There is over 100 volts on the driver's collector, which is located on the CRT socket board. The transistor isolates the high voltage from the IC. Later I'll show an application of this IC amplifier in the conversion process of a specific monitor.

Both circuit designs provide a comparison of these possibilities. As you can see, the analog IC version gives the most benefits with the least amount of work.

Concepts of Converting 25-kHz Monitors 3-7

You can get a fully functional VGA monitor for about $10. This is possible by converting surplus monitors so they will work with VGA. Have you ever seen old terminals (monitors) that were at one time used with mainframes for sale at ham and computer flea markets? These monitors work at 25 kHz and are limited to RGB video. At one time it was possible to connect them to super EGA cards, but now these cards are no longer available. A much better solution is to convert them to VGA. Make sure you check the model number before buying several of them. Some models are black and white and not as easily converted.

The whole concept of converting monitors took a positive turn when I discovered the similarity between a 25-kHz monitor and a 30-kHz monitor. Both were surplus IBM monitors with only a handful of differences. At first, I made several changes to convert from one frequency to the other, but with time I found that many are not required.

The model 2115 is designed to operate at 30 kHz, while the model 3179 is designed to operate at 25 kHz. Table 3-7 shows some of the differences.

Part #	IBM 3179 Old Part	IBM 2115 New Part	Notes
C504	223	472	
R509	1K	2.2K	
R510	1K	6.8K	
R511	150K	Remove	
R513	56K	22K	Horizontal hold range
R514	5.6K	4.7K	
R523	2.4K	4.7K	
C552	682 1.2kV	472 1.2kV	HV monitor circuit
R562	Jumper	1.5-ohm, 2-W	Filament voltage
C564	152 2kV	182 3kV	
C403	273	Remove	Stability improves

NOTES:

1. Some monitor differences are not listed because they are not relevant in the conversion process.
2. Remove the adjusting slug from the Horizontal Size control to attain maximum available horizontal width.
3. Adjust the power supply to a maximum value that still provides normal operation. This is typically up from 88 volts to a new setting of 98 volts.
4. I cannot accept responsibility for the proper or safe operation of a modified monitor. Safety precautions and procedures should be strictly adhered to in each and every design modification outlined.
5. This is not a complete list of all the changes that were made by the manufacturer.

Table 3-7: Differences between models 2115 and 3179

There are several monitor standards currently being used. Most older monitors are designed to be used at only one horizontal and vertical rate, whereas modern Super VGA monitors can handle several different horizontal and vertical frequencies.

Table 3-8 is a list of some of the most popular graphics standards and their scan rates. Notice that the vertical frequency varies very little from one standard to another. Only the horizontal section needs major modifications for a change from one standard to another.

Multiscan or multisync monitors are ones that operate at more than one scan rate. They work by varying several voltages and frequencies. Usually, a computer chip inside the monitor keeps track of what is needed, as well as the user preferences for each frequency.

Monitors designed to work at higher frequencies have a different deflection yoke design that gives them a much lower impedance. A practical limit is that a 25-kHz monitor can be converted up to 31 kHz. Any monitor that operates above 31 kHz can be converted down to 31 kHz. Therefore, a 15-kHz monitor cannot be converted up to 31 kHz. Converters are available however that will convert 31 kHz down to 15 kHz, so these can be used with such monitors. Two quite economical 31-kHz to 15-Hz converter designs are included in this book.

To convert to another fixed frequency, there are at least four areas that need to be modified. The horizontal hold is modified in order to increase the range of the horizontal frequency. A smaller resistor gives a higher frequency. The yoke capacitor is changed to keep the picture linear. This change is needed so that objects in the middle are the same width as they are at the sides. A smaller capacitor works better at a higher frequency. The capacitor connected between the horizontal output transistor's collector and ground will typically need to be changed to a smaller value. This shortens the retrace time so that the edges of the picture are not chopped off. The B+ is increased to attain a wider picture at the higher frequencies. Although a higher B+ is needed, the current actually decreases, and the total power remains approximately the same.

After the conversion, some optional circuits can be added. Optional features include automatic operation of the vertical size in different VGA modes or analog video amplifiers that maintain the integrity of color shades when using analog video.

Graphics Card	Horizontal	Vertical	Resolution
CGA	15.75 kHz	60 Hz	640 × 200
MGA	18 kHz	50 Hz	720 × 350
EGA (DOES CGA)	22 kHz	60 Hz	640 × 350
SEGA	25 kHz	60 Hz	640 × 400
VGA	31 kHz	60–70 Hz	640 × 480
SVGA	35 kHz	60–70 Hz	800 × 600
SVGA (noninterlaced)	48 kHz	60–75 Hz	1024 × 768

Table 3-8: Scan rates of popular graphics standards

IBM Model 3179 Conversion 3-8

Step-by-Step Conversion

1. Remove the cover. The top screws are behind snap-out covers, about 2 inches back from the front edge. The bottom screws are behind small "stick-on" plastic covers.
2. Clip off R511 (150K) and C403 (27000 pF). This gives wider horizontal and vertical ranges. They are located in the center of the main deflection board.
3. Add a 47K resistor across R513 (56K). Also add a 22K resistor across R514 (5.6K), which is located underneath the middle of the main deflection board. These changes increase the horizontal frequency.
4. Either make a 25-pin to 15-pin (high-density) adapter or remove the 25-pin plug, then solder a 15-pin (high-density) plug as a replacement. You can also make a homemade 25- to 9-pin adapter and then use a standard 9- to 15-pin NEC monitor type adapter that will fit onto the computer. Adapter pinouts that connect the monitor to a computer are listed in Table 3-9.
5. Turn the monitor on and adjust as follows:

   a. Turn up the screen control until video is displayed.
   b. Adjust the horizontal hold until the picture stabilizes.
   c. Adjust the vertical hold until it stops rolling.
   d. Adjust the vertical size and position as necessary.
   e. Turn the power supply (VR 81) up to 94 volts as measured at B3.

Signal Name	25-pin IBM 3179	9-pin NEC	15-pin (HD) VGA	Wire Color
Red	15	1	1	Yellow coax
Green	16	2	2	Brown coax
Blue	17	3	3	Blue coax
Horizontal	21	4	13	Blue
Vertical	22	5	14	Purple/pink
Ground	14	6-9	5-8, 10, 11	Shield braid

Table 3-9: Adapter pinouts to connect IBM model 3179 to a computer

Quantity	Part Number	Source
1	74LS86/74HC86	Mouser
1	.82 µF 400 V	Mouser
1	1-ohm, 2-watt resistor	Mouser
1	25-pin D-sub, male	Mouser
1	15-pin HD D-sub, male	Mouser

Note: Miscellaneous resistors are also needed.

Table 3-10: Parts sources

6. Remove the horizontal width coil slug (located on left side behind the horizontal output transistor). As an option, you can jumper across L554.
7. As an option, change the yoke capacitor (C560) from 1 to 0.82 µF to get a wider and more linear picture.
8. As an option, change jumper 79 to a 1-ohm, 2-watt resistor. This slightly reduces the filament voltage and it will help the picture tube last longer.
9. Use the circuit described earlier in this chapter on "Sync Phase Adapters" to correct the sync phase. This can be built on a small board and inserted between connector B1 and the deflection board. The 5 volts can be obtained from the video board.
10. Clip the 82-ohm resistors on the small video board. This gives some level of VGA compatibility.

Analog Video Amplifier for the 3179

The TTL video amplifier is normally an open collector TTL IC that drives a level shifting transistor. This in turn drives a 120-volt transistor that feeds the picture tube cathodes. In order to convert the video amplifier to analog, remove the TTL IC and replace it with an LM733 analog video amplifier. The negative input of the LM733 is used as a bias control to adjust the output voltage swing. Ideally, you need a gain of 10. Therefore, a 0.6-volt input will result in a 6-volt output. However, the amplifier can't drive to 6 volts output without clipping. By adjusting the bias, you can cause the output to go from –1 volt to +5 volts, and this eliminates the clipping problem.

If you can't find LM733s, three NE592s will do. You will need a 1K resistor from pin 4 to pin 11 in order to set the gain at 10 on the NE592. If there is a little

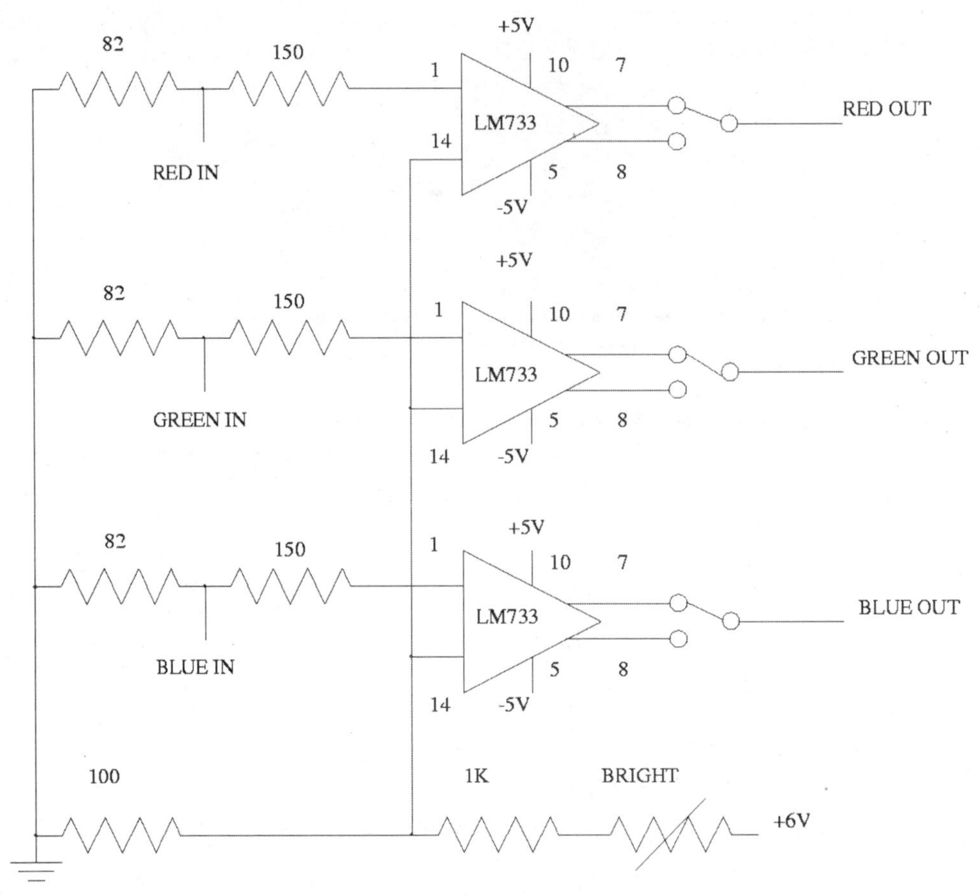

Figure 3-11: LM733 video amplifiers

smear in the picture, this can be eliminated by adjusting the gain controls that are found on the picture tube circuit board. The required power source of –5 volts is obtained from the vertical's –9 volts (at the junction of R428 and R433) and regulated by way of a 4- or 5-volt Zener.

Step-by-Step Video Conversion

1. Remove small video board and the following components:

   a. IC1302 (74LS38)

   b. R1301, R1303, R1305 (1K)

Quantity	Part Number	Source
3	LM733	MCM Electronics
3	150-ohm resistors	Radio Shack
3	82-ohm resistors	Radio Shack
1	100-ohm resistor	Radio Shack
1	1K resistor	Radio Shack

Table 3-11: Parts sources

c. Q1301, Q1303, Q1305 (C2901)
d. R1311, R1312, R1313 (330 ohm)
e. Q1311 (C1383)
f. R1321, R1322, R1323 (330 ohm)
g. R1315 (sub brightness)

2. Cut the runs that lead to the pads of the IC1302, solder three LM733 ICs one on top of the other connecting pins 5, 10, and 14 together, and then solder into IC1302s location.
3. Connect R, G, and B inputs found at the base of Q1301, Q1303, and Q1305 to pin 1 of the LM733s.
4. Connect R, G, and B outputs found at R1311, R1312, and R1313 to pin 7 of the LM733s.
5. Pin 14 on each of the LM733s should connect to a 1K resistor between Q1311's base and a 100-ohm resistor to ground. Q1311 was fed by the brightness control; now it sets the bias.
6. Pin 10 of the LM733s connects to 5 volts at C1302
7. Pin 5 of the LM733s connects to –9 volts at the junction of R428 and R433. This is found on the right side of the deflection board and connected via a 4- or 5-volt Zener to reduce the voltage to 5 volts.

IBM Model 5272 Conversion 3-9

This monitor is much like the 3179; however, the video section is quite different. I have not yet tried to convert the video to analog on this model, because the TTL level inputs seem to work exceptionally well for a basic VGA display.

Step-by-Step Conversion

1. Remove the cover using the same procedure as with the 3179. The top screws are behind snap-out covers about 2 inches back from the front edge. The bottom screws are behind small "stick-on" plastic covers.
2. Add an 18K resistor across the 5.1K resistor between pins 13 and 14 of the horizontal/vertical oscillator chip. This IC is located in the center of the deflection board. This increases the horizontal range.
3. Remove the horizontal width slug. This will provide a wider picture.
4. Remove the 82-ohm resistors that are located next to the video processor chip on the circuit board that plugs into the CRT. This gives compatibility with analog video because this monitor cannot be easily converted to operate with full analog video.
5. Rewire the input cable and connect it to a 15-pin connector or make an adapter. (Sorry, I don't have a pinout, but they are very similar, if not identical, to the 3719.)
6. Turn the monitor on and adjust the power supply to 97 volts. All other adjustments are identical to the 3719.
7. As an option, change the yoke capacitor from 1 to 0.82 μF. This gives better linearity and a wider picture.
8. As an option, replace C3412 in the power supply. This improves monitor start-up reliability.
9. When adding a sync phase circuit (from Section 3-4), the 5 volts can be obtained by way of pin 5 on plug 2C5. Pin 5 actually provides 9 volts. It can be connected to a series voltage-dropping 150-ohm resistor and then to a 5-volt Zener to ground to get a regulated 5 volts. Or simply use a 78M05 voltage regulator.

Note: Some 5272s have 43-ohm video resistors as described in step 4. In these models, change them to 100- or 150-ohm resistors. If you clip them, the video levels will be too high, and the picture will smear.

TAXAN Model 630 Conversion 3-10

This monitor has a connector similar to the one used with the IBM keyboard. A special adapter is made using a plug available from Radio Shack. Be careful,

Signal Name	Taxan 630	NEC (9-pin)	VGA (15-pin HD)
Red	2	1	1
Green	3	2	2
Blue	4	3	3
Horizontal	7	4	13
Vertical	8	5	14
Ground	6	6-9	5-8, 10, 11

Table 3-12: TAXAN model 630 adapter pins and connections

because the plug's pins do not go in the order of their appearance. The numbers are stamped on the plug in the following order: 6, 1, 4, 2, 5, 3, 7, and pin 8 is in the middle. The pins and connections are given in Table 3-12.

Step-by-Step Conversion

1. Make the adapter cable described in Table 3-12.
2. Add a 100K resistor across R703 (6.8K). This will extend the horizontal hold range. Then adjust the horizontal hold control all the way up.
3. Remove the 4 R/C combinations soldered on the bottom of the circuit board. (*Note:* They limit the video frequency response.)
4. Add jumpers across the 100-ohm resistors that are in series with the video signal (R103, 133, 163). This will increase the video sensitivity.
5. Turn VR 920 located on the power supply board all the way up. This increases the B+ to 122 volts.
6. Remove the horizontal width slug to get a wider picture.
7. Remove C721 located in the deflection area. There are actually two capacitors connected in parallel at C721. Separate them and solder the one marked 104 (0.1 μF) back into C721. This improves the horizontal linearity.
8. Remove C717 and C718, capacitors which are marked 103 (0.01) at 630 volts, and replace them with 0.0047 at 630 volts. This keeps characters from being chopped off at the right and left edges of the picture.
9. Remove R604 (4.3K) located behind the vertical hold and replace it with a 6.8K resistor. This will extend the vertical hold range. The sync

phase dip switch and vertical hold need to be adjusted for different vertical frequencies.

10. Adjust the focus control to the end of its range that provides the clearest picture.

This monitor has pull-up resistors on its inputs of 3.3K. This makes it more sensitive to the analog input, but may cause some noise. A better solution would be to remove the PROM video processor chip and replace it with an open collector output such as the LS TTL buffer. This will require removing most of the circuit around the PROM and adding jumpers to get the proper signals on the correct pins of the replacement chip.

IBM Models 65X2114 and 6405301 Conversion 3-11

These monitors are very similar in frequency to VGA's 31.5 kHz. Little modification is required to convert them to VGA compatibility.

The model 65X2114 is the same as a model 3179, but has a frequency already set at 30 kHz. All you need to do is adapt its connector to VGA and just adjust the horizontal hold control until you get a picture. You may want to adjust the horizontal size as well. Then, you can add the analog video amplifier and sync phase correction circuits that were discussed earlier in this chapter.

The model 6405301 has a frequency that is fixed at 30 kHz. In order to convert it to 31.5 kHz, change the resistor that is located between pins 10 and 13 of the TDA2582 chip from 27K to 25K. This resistor is located in the center of the main deflection board. Then make an adapter to convert to a VGA connector. Next, add the sync phase and analog video amplifier circuits just as you did with the 3179 monitor.

Mitsubishi and Princeton Monitor Conversion 3-12

Some other monitors are *much* easier to convert. Many have a 9-pin connector on the back similar to those found on some NEC monitors. They may also have an analog/TTL input switch. If that is the case, the conversion is easy. Just get a 9-pin to 15-pin adapter or cable and switch to analog video inputs. Some still need the sync phase correction circuit that was shown earlier with the 74LS86.

Signal	Mitsubishi (25-pin)	NEC (9-pin)	VGA (15-pin HD)
Red	2	1	1
Green	4	2	2
Blue	14	3	3
Horizontal	16	4	13
Vertical	17	5	14
Ground	1, 3, 5	6-9	5-8, 10, 11

Table 3-13: Mitsubishi adapter cable

Mitsubishi has such a monitor, but the analog input is on a 25-pin connector, not the usual 9-pin! This requires a special 25-pin to 15-pin adapter that's described in Table 3-13. (*Note:* I reverse-engineered this adapter, and it required several attempts before I was able to get it to work properly.)

Princeton also made a monitor that is very easy to convert. The model number was something like "Ultra 16." It also came with a 25-pin connector. I have seen only one of these models. Unfortunately, when I performed this conversion, I did not document the adapter.

The adapter is easy to make by trial and error, however. Use a voltmeter to find the grounded pins and connect them first. Try the colors one at a time until you get a red picture, then green, then blue. Then try the sync signals until the picture locks horizontally and vertically. The only dangers with this "trial and error" method of making adapter cables are to avoid shorting the signals from the VGA card to ground, and always to disconnect any pins on the monitor that go to a power source.

Another good idea is to use a junk VGA card when doing conversions.

64-kHz to 31-kHz Conversion 3-13

Now, we will convert high-frequency monitors down to 31 kHz. This is much more dangerous, because the high-voltage power supply can get out of control. This can lead to high-voltage arcing, damaging the fly back transformer, focus divider, and picture tube, and seriously injuring the person performing the conversion—you.

The popularity of Super VGA graphics cards has grown, and they now sell for under $50. They can scan up to and beyond 65 kHz, making this conversion not even necessary. One popular card that works well with these 65-kHz monitors is the "Diamond Stealth 3D-2000." However, if you're like me, you prefer troubleshooting at the DOS prompt. One of these days, though, I'll convert totally to Windows and leave my monitor at 1280 × 1024 resolution mode at all times. Another solution is to use a monitor switch to between the 65-kHz monitor and a small normal 31.5-kHz type VGA monitor.

The person making these modifications must accept all responsibility for safety. Levels of high voltage must be observed at all times with a high-voltage probe. The author cannot be responsible for the safe operation of a modified monitor.

We will discuss the conversion of a 64-kHz Hitachi monitor model HM-4119. I have converted a dozen or more of these, some years ago, and they all still work fine today. The principles shown in this model apply to most others as well. Everything is modified oppositely to the way a monitor is converted up to 31 kHz. The horizontal hold must be increased in value, the yoke capacitor must be increased in value, and the power supply must be substantially reduced, by as much as one-half.

First the horizontal hold is covered with epoxy to keep it in place and prevent tampering. It needs to be removed and replaced with a 4.7K variable resistor. Also, in order to increase the horizontal hold's range, add a 10K resistor in series with it.

Secondly, reduce the B+ to 60 volts by disconnecting JP 234. This is located just under the front left corner of the pincushion transformer (T-201). The power to the horizontal section is then supplied by a tap into the 60-volt source. A jumper is connected from the loop between pins 3 and 4 of the yoke plug (D02) then to a tap connected to the video section (pin 3 of V04 on the video board). This tap can be made in the wire at a point that is closer to the yoke plug, by tracing the wire back from V04.

In order to establish good voltage regulation and help stabilize the picture, remove R114 (82K) near the voltage regulator adjustment located on the right front side of the main circuit board. Run a 27K resistor at the point where R114 was connected to VR 100 (B+ adjustment) to the 60-volt supply at C118's positive end. This is easiest to do underneath the circuit board. Next, remove C117

(.047 μF) and connect it to the same point to which it was connected at VR 100; the other end also connects to the 60-volt supply.

Third, the linearity or yoke capacitor needs to be 1.5 μF at 250 volts, and it cannot be an electrolytic type. It is connected from pin 5 of T-201 to ground and it can be connected across C230 underneath the chassis.

The fourth step is to reduce the high voltage. Without this change, the high voltage will rise dramatically to 28 kV or more. To cure the excessive high voltage, add a 0.0015 μF, 1600-volt capacitor across C204 or C205. This is the same as connecting it from the collector of the horizontal output transistor to ground.

The monitor will need a 15-pin to BNC connector in order to adapt it to the VGA card. You can make one yourself or buy one from LYBEN at (810)268-8100. BG Micro may also have these adapter cables available.

Don't be surprised if the picture rolls vertically—the sync phase is not yet corrected. Try to adjust the horizontal and vertical hold as closely as possible, in order to get a test picture on the screen. Adjust the horizontal position on the upper right-hand corner of the video board in order to center the picture on the screen.

The seventh step is to add a "sync phase processor" to correct the vertical phase and stop the picture from rolling. The circuit that is to be added uses the 74LS86, discussed earlier in this chapter. To use the composite sync output, set it up with the two jumpers moved to the opposite positions from what is shown in the schematic.

Frequency	B+ Volts	H Hold	Yoke Cap	Retrace Cap
64 kHz	120V	2K		
31 kHz	60V	15K	Add 1.5 μF	Add 0.0015

NOTES:

1. You will need to adjust the horizontal position trimmer for each frequency.
2. The manual specifies the high voltage as 27 kV at 64 kHz, and 25 kV at 48 kHz. The converted monitor should have less than 25 kV.
3. If the monitor fails to start, change R208 from 6.8K to 12K in the power supply circuit.
4. On the model 4129, connect a 3.3K resistor in parallel with the 3.3K resistor (R21) on the power supply regulation board.

Table 3-14: Parts changes to convert from 64 kHz to 31 kHz

Quantity	Part Number	Source
1	74LS86/74HC86	Mouser
1	1.5 µF 250-volt metal	Mouser
The 0.0015 can be obtained by connecting the following two capacitors in parallel.		
1	470-pF, 2000-volt ceramic	Mouser
1	1000-pF, 2000-volt ceramic	Mouser
1	15-pin high-density to 5 BNC cables adapter	Lyben

Table 3-15: Parts sources

The circuit can be situated internally or externally, but it will need 5 volts to operate. This is available internally via pin 16 on the 74123s, which are located on the video board just above the sync input jacks. The sync output is inverted via the 4TH gate, and this will match the HM-4119's composite sync input's phase requirement.

For easy adjustment of the vertical size, a 100K variable resistor can be added on the front panel of the monitor. You can mount this control above the contrast control. One end connects to ground and the wiper connects to the vertical size control located on the back panel. Be sure to use a shielded wire or the picture will shake. This facilitates manual adjustment of the vertical size. Automatic adjustment may be tied into some mode control relays, and separate trimmers can then be used for each frequency.

These are the apparent values required for operation at different frequencies. A relay can be set up to select between the desired modes of operation. The horizontal retrace capacitor needs its own relay because of the high voltage on it. Always default to the lowest frequency and the lowest power supply voltage.

IBM PS-2 Model 8525 Conversion 3-14

The model 8525 was another failed attempt at a network computer. This monitor includes a 286 computer, but no hard drive. This design keeps resurfacing. The idea is to exclude the disk drives and control everything from the server. Somehow this is *supposed* to save you thousands of dollars on a computer. In my current position I work as a network administrator, computer repairman, and program writer, and in my spare time I've installed and maintained the telephone system, which has more than 130 lines. We spend less than $10,000 per year on about 60 computers.

Using computers that are one step behind the "state of the art" saves us a substantial amount of money, and these machines are much easier to maintain. These computers include hard and floppy drives and still cost less than "network computers." Yet they do not put as much of a strain on the network. Forget network computers, *please*!

The model 8525 has a monitor and power supply located in the unit's upper half. The bottom comes off by removing two screws located on the back and then sliding it up and out. All connectors can then be disconnected. Those that carry the video to the monitor can be identified by the black shielded coax wires. I didn't know the correct pin numbering system, but by using reverse engineering I was able to find this information. The simplest method is to get a 15-pin high-density male plug and a 14-pin header connector and make an adapter. Then use a 15-pin VGA extension cord to connect to the computer.

The header pinout in Table 3-16 is listed with the header's pins aimed at you.

This is the best method for making this adapter. Solder the header pins 1, 2, and 3 to the 15-pin connector pins 1, 2, and 3. If this is properly aligned, the header pin 5 should also align with pin 5 of the 15-pin connector. Snip off pin 4. The left row of header pins should hang to the side of the 15-pin connector. Run a jumper from pins 8 to 11 of the header to pins 6 to 8 of the 15-pin connector. This jumper forms a 1/4-inch "U" and can be made of 20-gauge copper wire. Then solder bridge pins 5 and 10, and pins 6 and 11, on the 15-pin connector. Then add the horizontal and vertical jumpers using insulated wires. Next, screw the adapter into the knock-out hole in the back of the monitor's cabinet. Use

VGA	Signal	Header	Header	Signal	VGA
6	Ground	8	1	Red	1
7	Ground	9	2	Green	2
8	Ground	10	3	Blue	3
11	Ground	11	4	Ground	
13	Horizontal	12	5	Ground	5, 10
14	Vertical	13	6	???	
	???	14	7	Key	

Table 3-16: Header pinout

washers to fill the gaps. Then plug the video cable into the header pins. Add the VGA extension cord and you should have a working monitor.

Decision Data DSM-5235 3-15

This monitor makes a very nice VGA monitor once it is converted. It is normally part of a diskless workstation like the IBM 8525. I have found this monitor being thrown out in the garbage when it was only 2 years old. It is a Super VGA monitor capable of 800 by 600 and even 1024 by 768 resolution.

The only problems with this monitor are that it has a 15-pin plug like an Apple monitor, and the vertical needs to be inverted for it to work as a VGA monitor. There is no horizontal width control on the front panel, and the picture gets narrower in the higher resolution modes. An extra power supply is located inside to power the terminal. It is best to remove it, to avoid accidentally connecting the power pins to your VGA card.

To convert this monitor to VGA, do the following:

1. Remove the base unit—it is an 80186 workstation terminal. It can be kept for a tilt and swivel base.
2. Remove the screws and remove the rear cover.
3. Remove the power supply. It is only needed for the terminal portion. This power supply is a circuit board about 4 × 2 inches on the right side of the chassis. It has three plugs on it—P171, P172, and P173—and two mounting screws.
4. There is a 74LS86 phase correction circuit already built into the monitor,

15-pin Connector	Signal	15-pin HD VGA Connector
1	Ground	6
2	Ground	7 and 8
3	Red	1
4	Green	2
5	Blue	3
11	Horizontal	13
12	Vertical	14
13, 15	Ground	5, 10, and 11

Table 3-17: Pin connections for 15-pin male to 15-pin HD male adapter

but it doesn't work correctly for the vertical. To correct the vertical phase, add a 2N3904 transistor as follows:

a. Cut the run between P401B pin 5 and J172.
b. Add a 470-ohm resistor from the base of the 2N3904 to P401B pin 5.
c. Solder the 2N3904's emitter to ground at the 74LS123 pin 1, 6, 8, 9 or 14.
d. Solder the 2N3904's collector to J172.
e. Solder a 2.2K resistor from J172 to 5 volts at pin 11 or 16 of the 74LS123.

5. Make a 15-pin male to 15-pin HD male adapter with the pin connections given in Table 3-17.

Apple Multiple Scan Adapter 3-16

This monitor needs only a 15-pin female to 15-pin HD male VGA-type plug. It is very Super VGA compatible and is capable of 800 by 600 as well as 1024 by 768 without any loss of picture width. It also features digital controls as well as memory to remember your preferences.

Not all Apple monitors can use the same VGA adapter. I had three Apple-type adapters sitting around and tried each of them. One of the adapters fried my VGA card! Fortunately, I use an old VGA card when making and testing adapters.

Some Apple-type monitors use composite sync on pin 3 of their 15-pin connector. If this is the case, refer back to the 74HC86 sync phase adapter. It can be set up to provide the needed composite sync for these Apple monitors.

15-pin Connector	Signal	15-pin HD VGA Connector
1	Ground	6
2	Red	1
5	Green	2
6	Ground	7
9	Blue	3
11	Ground	10, 11
12	Vertical	14
13	Ground	8
15	Horizontal	13

Table 3-18: Pin connections for 15-pin female to 15-pin HD male adapter

Common Monitor Repairs 3-17

With time and experience, it seems that when you see a monitor at a distance you can tell the make, model, and what is most likely wrong with it. So, to share my experience with others, I've started a list of common problems. Most of these I've seen more than once or twice.

Make: Various manufacturers

Problem: Poor focus

Solution: Remove glue and adjust the focus control located on the high-voltage transformer. You might also want to drill a small hole in the cabinet for future adjustment. Eventually you may need to replace the fly-back transformer or toss the monitor.

Make: Various manufacturers

Problem: Vertical or horizontal roll, or one color changes or is missing

Solution: Replace the video cable. Very often it gets bent too tightly and develops internal shorts. You can use the cable from monitors that were not easy to fix and were then discarded. On several occasions I have successfully resoldered the 15-pin high-density plugs, but don't recommend it.

Make: Packard Bell (this happens on a few other models as well)

Problem: It seems to be running but there is no picture, or the color hue shifts slightly

Solution: Resolder the CRT connector. You will need to remove all shielding around the CRT socket to get to it. On some models you can just remove the rear cover of the shielding.

Make: Older Compaq 14-inch

Problem: Vertical flips, jumps, and rolls

Solution: This usually gets worse as the monitor warms up. Replace the TDA1170 vertical IC chip. The problem is typically from overheating. You may want to make sure objects aren't placed on or blocking the monitor's vents.

Make: IBM PS/Value Point

Problem: Power supply makes a ticking noise

Solution: Replace the horizontal output transistor.

MAKE: IBM G-50

PROBLEM: Power supply not working

SOLUTION: Replace 1000-μF, 25-volt capacitor in power supply section.

MAKE: Gateway 2000 Crystalscan (PMV-1448 or equivalent)

PROBLEM: Power supply won't start

SOLUTION: Replace the 200-μF, 25-volt capacitor in the power supply.

PROBLEM: Shaky picture

SOLUTION: Resolder all connections on the deflection board.

PROBLEM: Vertically short

SOLUTION 1: Replace R203, a 390K resistor in the vertical section of the deflection board.

SOLUTION 2: Add a 100K resistor from W202 (vertical size) to ground.

MODEL: Apple Multiple Scan 15 display (Model M2943)

PROBLEM: Color shifts

SOLUTION: Remove the shielding around the CRT socket and resolder the socket pins, as well as the pins of the transistors that are mounted to heat sinks.

MODEL: Kehtron (Model CX-4158)

PROBLEM: Rolling horizontally or vertically in some modes

SOLUTION: Replace IC part number TDA9103 or add a 74HC86 adapter to correct the phase.

PROBLEM: Forgets preferences whenever mode changes or power is turned off

SOLUTION: Replace serial EEPROM 24C04.

MODEL: Samtron 14 inch SVGA

PROBLEM: Vertical short, line through picture

SOLUTION: Replace the vertical IC, part number TDA8351. Turn the power supply down to 70 volts if needed.

MODEL: NEC 14-inch monitor

PROBLEM: Poor picture or two or three stripes vertically through picture

SOLUTION: Replace three 2.2-μF, 50-volt capacitors on CRT socket assembly next to video processor chip.

General Troubleshooting—All Monitors

Always check the power supply. If the fuse is blown, check for shorts. The horizontal output transistor can be checked by metering its collector to ground, or by metering the horizontal yoke to ground. Check all power supply outputs for shorts to ground. The CRT filament will look like a short, so unplug the CRT for testing that circuit. If necessary, disconnect the horizontal output and see if the power supply starts up. Never run a power supply without a load—always leave something online.

Look for leaky or "expanding" capacitors. They are the second most common cause of monitor failures. I've opened up one monitor and found a loose cap rattling around inside. It had expanded so much that it unsoldered itself! Many faulty caps will change in value with temperature. Sometimes I replace every capacitor in the faulty section. Another trick is to use "freeze spray." Freeze the capacitors in the faulty section one at a time, until you find the one causing the problem.

Check fusible resistors and power resistors. Sometimes they go bad but don't show it. With power off, meter them for the resistance value marked on them. There is an Impression 14-inch monitor that won't start when a fusible resistor becomes bad in the power supply.

About half of the monitors I get can be fixed in a few minutes. Others can be tough, if not impossible, to fix. Generally, if it cannot be fixed in an hour, strip off a few parts and discard it.

Another place to look for monitor repair advice is on the World Wide Web. There are several sites that offer free technical support, but the most helpful ones require an access charge.

Chapter 4

Conclusion

In this chapter you will discover:

This book contains a compendium of projects completed over several years. Many have been a lot of fun, and some never worked as well as I would have liked. One of the "duds" was a VGA to 64-kHz converter. Without multilayer circuit boards, the noise generated at the required frequencies made a mess of the resulting picture. The printer port mixer distorts too easily and has too much noise for professional use. The printer port EPROM verifier should have been able to program EPROMs as well. I'm sure that someday all these problems will be resolved, one way or another.

Some projects I keep revisiting and expanding. This seems to happen just when I thought the project had reached the end of its useful life. The EPROM copier is an example of such a device. It has been around for many years and keeps getting expanded to accommodate bigger EPROMs.

I hope that this book will be a source of help and inspiration that will get others started on the adventure of making their own digital and computer projects. Once started, you can easily become addicted to the concept "I could do that a whole lot more easily and cheaply."

Many people are predicting the "demise" of the hobbyist. Their reasoning is that assembled devices can often be purchased from overseas sources that are a lot cheaper. If something can be made, however, and is not readily available, the opportunity to make it yourself is there. Part of the problem is that "old-timers" have lost the inspiration and spend time on projects of little value. I hope that others like myself will bring new life into making digital and computer hobby projects that are interesting, fun, and useful.

To get started in making your own digital and computer projects, you need to research and develop a list of some ideas for devices or inventions that are of interest to you. Then, find a design that's simple enough that building it is practical. That takes practice, and you'll need to start somewhere. Try your hand at some of the projects in this book.

Some of you may find better or more cost-effective ways of doing what I have done. You may write faster or shorter code to get the job done. As one who loves to learn and share information, I would be delighted to hear from you. Good luck—and enjoy!

Bibliography 4-1

Dallas Semiconductor. *1992–1993 Product Data Book.*

Motorola Semiconductor. *Master Selection Guide.* Motorola Inc., 1992.

Motorola Semiconductor. *MC68HC11F1 Technical Data.* Motorola Inc., 1993.

National Semiconductor Corporation. *CMOS Databook*, 1981 edition.

National Semiconductor Corporation. *Linear Applications Handbook*, 1977 edition.

National Semiconductor Corporation. *Linear Databook*, 1980 edition.

National Semiconductor Corporation. *Memory Databook*, 1980 edition.

National Semiconductor Corporation. *Memory Databook*, 1990 edition.

National Semiconductor Corporation. *MOS Memory Databook*, 1984 edition.

NEC Electronics Inc. *Memory Products Data Book*, 1988.

Rosch, Winn L. *The Winn Rosch Hardware Bible*, 1989.

Texas Instruments Inc. *The TTL Data Book for Design Engineers*, second edition, 1981.

Parts Sources 4-2

MCM Electronics

650 Congress Park Dr.
Centerville, OH 45459
1-800-543-4330

MCM is possibly my number one supplier of parts, etc. They carry just about everything. They are a great source for semiconductors of all types, especially original components for monitors.

Digi-Key Corporation

701 Brooks Ave. South
Thief River Falls, MN 56701
1-800-344-4539

Digi-Key carries the Maxim series of IC's, PIC controllers, and much more.

Jameco Electronics

1355 Shoreway Road
Belmont, CA 94002
1-800-831-4242

Jameco carries some kits, lots of parts, memory chips, 8052AH basic CPUs, etc. They will provide specification sheets for almost every product they sell for just $1 more. They also have a selection of IC manuals available.

BG Micro

P.O. Box 280298
Dallas, TX 75228
1-800-276-2206

BG Micro carries a lot of surplus parts, such as LM-34 temperature sensors and opti-couplers, and all are at excellent prices.

Harbor Freight Tools

3491 Mission Oaks Blvd.
Camarillo, CA 93011
1-800-423-2567

They carry numerous tools—drill presses, needle-nose pliers assortments, drill bits, screwdrivers, etc.

Mouser Electronics

958 N. Main
Mansfield, TX 76063
1-800-346-6873

They carry IC sockets, TTL IC's, plastic project boxes, etc.

Newark Electronics

Chicago IL 60640
1-800-298-3133

Newark carries hard-to-find ICs such as memory, microprocessors, and analog-to-digital converters. To my knowledge they carry more ICs than any other supplier. They use area sales representatives, so call to find out where to place an order.

Marshall Industries

9320 Telstar Avenue
El Monte, CA 91731

They are the only source that I know of for the NEC video line buffers used in VGA-to-CGA converters.

Radio Shack

Fort Worth, TX 76102
1-800-843-7422

Radio Shack carries some ICs, numerous project boxes, switches, LEDs, lights, and light sockets. They are my main source for printed circuit board etchant. They also sell an assortment of resistors and capacitors at very good prices.

New Micros Inc.

1601 Chalk Hill Rd.
Dallas, TX 75212
(214)339-2204

They carry lots of parts, assemblies, and sockets that work with 68HC11's.

Lyben

(810)268-8100

They are a possible source for 15-pin to 5-BNC cable adapters.

HSC Electronics

1-800-442-5833

They also are a possible source for 15-pin to 5-BNC cable adapters.

IMSI

1895 Francisco Blvd. East
San Rafael, CA 94901-5506

They sell Turbocad, the CAD program I used for all of the schematics in this book.

Index